ISOFLAVONES: A RACE AFTER THE RESCUE OF THE AGING HIPPOCAMPUS

BIOCHEMISTRY RESEARCH TRENDS

Additional books in this series can be found on Nova's website
under the Series tab.

Additional E-books in this series can be found on Nova's website
under the E-book tab.

BIOCHEMISTRY RESEARCH TRENDS

ISOFLAVONES: A RACE AFTER THE RESCUE OF THE AGING HIPPOCAMPUS

CATHERINE BENNETAU-PELISSERO
KHALID JAMALI
AND
ALINE MARIGHETTO

Nova Biomedical Books
New York

Copyright © 2011 by Nova Science Publishers, Inc.

For permission to use material from this book please contact us:
Telephone 631-231-7269; Fax 631-231-8175
Web Site: http://www.novapublishers.com

NOTICE TO THE READER

The Publisher has taken reasonable care in the preparation of this book, but makes no expressed or implied warranty of any kind and assumes no responsibility for any errors or omissions. No liability is assumed for incidental or consequential damages in connection with or arising out of information contained in this book. The Publisher shall not be liable for any special, consequential, or exemplary damages resulting, in whole or in part, from the readers' use of, or reliance upon, this material. Any parts of this book based on government reports are so indicated and copyright is claimed for those parts to the extent applicable to compilations of such works.

Independent verification should be sought for any data, advice or recommendations contained in this book. In addition, no responsibility is assumed by the publisher for any injury and/or damage to persons or property arising from any methods, products, instructions, ideas or otherwise contained in this publication.

This publication is designed to provide accurate and authoritative information with regard to the subject matter covered herein. It is sold with the clear understanding that the Publisher is not engaged in rendering legal or any other professional services. If legal or any other expert assistance is required, the services of a competent person should be sought. FROM A DECLARATION OF PARTICIPANTS JOINTLY ADOPTED BY A COMMITTEE OF THE AMERICAN BAR ASSOCIATION AND A COMMITTEE OF PUBLISHERS.

Additional color graphics may be available in the e-book version of this book.

Library of Congress Cataloging-in-Publication Data

Bennetau-Pelissero, Catherine.
  Isoflavones : a race after the rescue of the ageing hippocampus /
Catherine Bennetau-Pelissero, Khalid Jamali, and Aline Marighetto.
      p. ; cm.
  Includes bibliographical references and index.
  ISBN 978-1-61728-751-0 (softcover)
  1. Isoflavones--Physiological effect. 2. Hippocampus (Brain) I. Jamali,
Khalid. II. Marighetto, Aline. III. Title.
  [DNLM: 1. Isoflavones--therapeutic use. 2. Brain--drug effects. 3.
Hippocampus--drug effects. QU 220]
  QP671.F52B46 2010
  615'.321--dc22
                          2010034035

Published by Nova Science Publishers, Inc. † New York

Contents

Preface   **vii**

Abbreviations   **ix**

Chapter I   Introduction   **1**

Chapter II   Dietary Isoflavones   **3**

Chapter III   Estrogen Receptors   **7**

Chapter IV   Brain Signalization by Estrogens
                     and Estrogen Receptors   **13**

Chapter V   Natures of Aging   **23**

Chapter VI   Pharmacological Effects of Isoflavones
                     on the Hippocampal Formation   **29**

Chapter VII   Pharmacological Effects of Isoflavones
                     in Non-Hippocampal Cells   **41**

Chapter VIII   Physio-Molecular Impact of Isoflavones
                     on Brain Physiology   **47**

Chapter IX   Effects of Dietary/Supplemental
                     Isoflavones on Cognitive Processes   **51**

Chapter X   Concluding Remarks   **65**

References   **71**

Index   **95**

Preface

Isoflavones are naturally occurring molecules often found in significant amounts in many leguminous, that include some edible plants. Isoflavones exhibit similarities in their chemical structure with steroidal estrogens. This allows them to trigger some of the endogenous physiocellular signalization processes of estrogens that are known to occur naturally within living organisms. In fact, isoflavones mediate numerous health effects mainly related to their estrogenic and/or anti-estrogenic actions. Some effects are the result of estrogen receptor-mediated physiological phenomenona such as those linked to the hypothalamo-hypophyso-gonadal axis. Moreover, because the expression of estrogen receptors is highly enriched in neural tissues, it was not surprising to discover that isoflavones were potent neurotrophic and neuroprotective compounds as it is the case for estrogens. Functional estrogen receptor proteins are also expressed in some central nervous system territories, including the hippocampus and the cerebral cortex, that are well thought to be devoted to higher cognitive brain functions such as memory function. Particularly, the hippocampus expresses both nuclear and membrane estrogen receptors known to mediate local physiocellular alterations of the neuronal phenotype, namely dendritic spine development, synaptic plasticity and modulation of the Ca^{++} trafficking rate between different compartments of the cell. These physiocellular effects are thought to be part of the adaptive chemical and architectural changes sustaining some processes of the mnemonic information storage. This provides an *in vitro* animal review and clinical data, to highlight the impact of isoflavones on the physiocellular features of hippocampal formation plasticity; and, therefore, on the hippocampus-dependent cognitive brain function. In summary, isoflavones may not always exhibit all the mechanistic effects of estrogens; but do,

viii   Catherine Bennetau-Pelissero, Khalid Jamali and Aline Marighetto

however, display some their own specific mechanisms of signalization. Nevertheless, the ongoing search efforts for neuroprotective selective estrogen receptor-modulating molecules that do not trigger the adverse side-effects of estrogens still deserve continuous investigation. More specifically, the putative neuronal survival-promoting and promnesic features of isoflavones need further exploration to get full acceptance as a beneficial natural substance against age-triggered alterations of general health including hippocampal-dependent memory decline.

Abbreviations

[35S]-GTPγ-S	sulphate 35-radio-labeled guanosine-triphosphate gamma-stimulating
[125I]estrogen	iode 125-radio-labeled estrogen
[Ca++]i	intra-cellular Ca++ concentration
α7nAchR	α7 nicotinic acetylcholine receptor
α7nAchRs	α7nicotinic acetylcholine receptors
5HT	5-hydroxy-tryptamine (serotonin)
AChR	acetylcholine receptor
AChRs	acetylcholine receptors
ACh	acetylcholine
AChE	acetylcholine esterase
AD	Alzheimer's disease
ADP	adenosine-di-phosphate
ATP	adenosine-tri-phosphate
AMPA	α-amino-3-hydroxyl-5-methyl-4-isoxazolepropionate
AMPAR	α-amino-3-hydroxyl-5-methyl-4-isoxazolepropionate receptor
AS-ODN	antisense oligodeoxynucleotide
AVP	arginin-vasopressin
AVPV	anteroventral periventricular nucleus
BDNF	brain-derived neurotrophic factor
b.w.	body weight
Ca++	calcium
CA1/2/3/4	cornu ammonis fields 1/2/3/4
cAMP	cyclic-adenosine mono-phosphate
cGMP	cyclic-guanosine-mono-phosphate

ChAT	choline acetyl transferase
ChAT-IR	choline acetyl transferase-immunoreactivity
CNG	cyclic nucleotide-gated
CSF	cerebro-spinal fluid
DA	opamine
DG	entate gyrus
DMP	delayed matching-to-place
DNA	deoxyribonucleic acid
DRG	dorsal root ganglion
E2	17-β-estradiol
EB	estradiol benzoate
ED	embryonic day
ER	estrogen receptor
ERα	estrogen receptor α
ERα-IR	estrogen receptor α-immunoreactivity
ERβ	estrogen receptor β
ERβ-IR	estrogen receptor β-immunoreactivity
ERK	extracellular signal-regulated kinase
ERK1/2	extracellular signal-regulated kinase 1/2
GABA	γ-amino-butyric-acid
GABA-A	γ-amino-butyric acid-receptor A
GDP	guanosine-di-phosphate
GFAP	glial fibrillary acidic protein
GAD67	glutamic acid decarboxylase-67
GAP43	growth cone-associated protein-43
GSK-3β	glycogen synthase kinase-3β
GPCR	G protein-coupled receptor
GPCRs	G protein-coupled receptors
GPR30	G protein-coupled estrogen receptor
GPR30-IR	G protein-coupled estrogen receptor-immunoreactivity
hERα	human estrogen receptor α
hERβ	human estrogen receptor β
IGF-1	insulin-like growth factor-1
i.p.	intraperitoneal
K+	potassium
KA	kainic acid
KCl	potassium chloride
LDH	lactate dehydrogenase

LTP	long-term potentiation
MAP-2	microtubule-associated protein 2
MAPK	mitogen-activated protein kinase
mGluR	metabotropic glutamate receptor
mRNA	messenger ribonucleic acid
MTT	3-[4,5-di-Methylt-Thiazol-2-yl]-2,5-diphenyl Tetrazolium bromide
Na+	sodium
nAchR	nicotinic acetylcholine receptor
NGF	nerve growth factor
NMDA	N-methyl-D-aspartate
OVX	ovariectomy
Oxyt	oxytocin
PCR	polymerase chain reaction
PELP1	praline/glutamic acid/leucine-rich protein-1
Phyto-600	soy phytoestrogen-rich diet containing 600 µg/kg of diet
PI3K	phosphate-inositol-3 kinase
PND	post-natal day
pp60c-src(+)	cellular counterpart of Rous sarcoma virus protein tyrosine kinase
PSD95	postsynaptic density protein-95
PTK	protein tyrosine kinase
RNA	ribonucleic acid
RT	reverse transcriptase
SERCA	sarco-endoplasmic Ca++ ATPase
SOCCs	store-operated Ca++ channels
TrK	tyrosine kinase
VDCC	voltage-dependent calcium channel

Chapter I

Introduction

Estrogens belong to a family of signaling molecules, called sex steroid hormones that are endogenously produced in animals. Multiple sites of production of endogenous estrogens have been identified in mammals, including the gonads (Mendelson *et al.*, 2005), the adrenal glands (Pelletier *et al.*, 1992), the adipose tissue (Thijssen *et al.*, 1993) and the central nervous system (Garcia-Segura, 2008), to name a few. It is well established that estradiol (E_2) derives from cholesterol through several metabolic steps where the rate-limiting, ultimate step is represented by the enzymatic aromatization of testosterone to estradiol (Ohlsson *et al.*, 2009). This aromatization occurs in all the tissues previously mentioned, and more specifically, in both cerebral neuronal cells and glial cells (Garcia-Segura, 2008). In the mammalian females, the ovaries represent the major site of synthesis and endocrine release of estrogens. Through blood circulation, the ovaries are capable of supplying the rest of the organism with estrogens, during the span of life. Within the family of estrogens, estradiol represents one of the most biologically active molecules. Besides, phytoestrogens are plant derived molecules (Albertazzi and Purdie, 2008; Cornwell *et al.*, 2004; Fitzpatrick, 2003) with structural characteristics resembling those of E_2. They exhibit two hydroxyl groups on the opposite sides of the molecule at a distance of 10 Å forming an angle similar to the one that exists between the two hydroxyl groups of estradiol (Katzenellenbogen *et al.*, 1980). Five major classes of phytoestrogens have been identified: Isoflavonoids, flavonoids, stilbenes, lignans, and coumestans. Isoflavones are a sub-class of isoflavonoids that includes genistein, daidzein, glycitein, biochanine-A, and formononetin (Figure 1). Isoflavones can be found in significant amount in some food plants such as soy, kudzu, or clover

(Setchell, 2001). Under their natural form, unprocessed isoflavones occur mainly as glucoside (Figure 1), malonyl or acetyl molecules (for example, in soybean hypocotyls) (Anderson & Wolf, 1995) that are chemically bound to a protein matrix. When ingested, these isoflavone-glucosides are subjected to enzymatic deconjugation in the gut tract. This leads to the release of the aglycone moiety being the active molecules. Other enzymatic transformations end up with the production of even more powerful metabolites. It is the case, for example, for equol (Figure 1), which can arise from formononetin or daidzein as a result of the action of gut tract-harbored bacterial flora (Setchell et al., 2002). Further enzymatic transformations are intended to facilitate the transport and/or the excretion of assimilated isoflavones. They take place in the enterocytes, hepatocytes or kidney cells and lead to the formation of glucuronids and sulfate conjugates (Rowland et al., 2003).

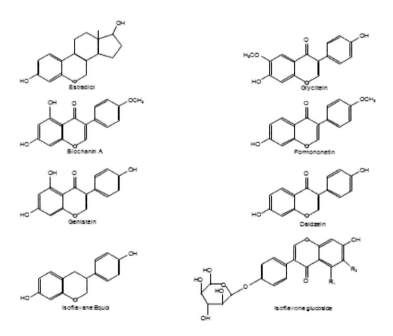

Figure 1. Molecular structures of estradiol, isoflavone, isoflavane, and isoflavone-glucoside species. Notice the close resemblance of isoflavones (glycitein, biochanin A, formononetin, genistein, daidzein) and the isoflavone metabolite, isoflavane, equol to estradiol. Dietary isoflavones occur in vegetables mainly as glucoside molecules: daidzin ($R_1=R_2=H$), genistin ($R_1=OH$; $R_2=H$) and glycitin ($R_1=H$; $R_2=OCH_3$).

Chapter II

Dietary Isoflavones

Due to their structural resemblance to the active steroid, E_2 (Figure 1), which allows them to display a relatively high affinity for estradiol receptors (ERs) (Belcher and Zsarnovsky, 2001; Kuiper *et al.,* 1997; 1998), isoflavones are able to trigger directly some of the endogenous physiocellular signalization processes of estrogens that are known to occur naturally *in vivo.* Indeed, a large set of data did establish that isoflavones were capable of influencing welfare and of relieving some age-triggered health alterations in humans, such as menopausal symptoms which include mild memory dysfunctions, advancing age-related mood state dysregulations (depression), bone metabolic defects (osteoporosis) and vasomotor-originating thermogenic discomforts (hot flushes) (Setchell and Cassidy, 1999; Cornwell *et al.,* 2004; Ferrari 2004; Kreijkamp-Kaspers *et al.,* 2004; Geller and Studee, 2006; Zhao and Brinton 2007b; Hooper *et al.,* 2009). In this context, isoflavones were also shown to affect numerous physiological parameters in animal models (Clarkson *et al.,* 1995; Lephart *et al.,* 2004; Usui *et al,.* 2006; Xiao, 2008) some of which concern neurochemical and hormonal interactions within the hypothalamo-pituitary-gonadal axis (Lephart *et al.,* 2005; Setchell and Cassidy, 1999). Empirically the estrogenic impact of isoflavones came to light as adverse anti-reproductive and pro-abortive effects in sheep grazing on red or subterranean clover pastures (Bennetts *et al.,* 1946; Braden *et al.,* 1967). Subsequently, the resulting studies have established that the incriminated components were isoflavones from clover also present in soy and kudzu, and that these molecules were able to distort the normal neurochemical/hormonal signalization within the hypothalamo-hypophyso-gonadal axis (Findlay *et al.,* 1973). These occurrences were made possible, most likely, because isoflavones, to some extent, are able to replace faded E_2 within the molecular interactions governing estrogen signalization (Kuiper *et al.*, 1997; 1998). Isoflavones are also capable of dislodging existing E_2 from the signalizing molecular tangle (Thompson *et al.*, 1984; Kuiper *et al.*, 1997; 1998) and, thereof, functionally sliding into specific intra-cellular signalization pathways taking place in various cell/tissue/organ systems (Watson *et al.*, 2007). Isoflavone-triggered, E_2-mimicking signalization does also occur in the higher central nervous system (Belcher and Zsarnovsky, 2001).). Another interesting aspect of isoflavones is reflected by their ability to modulate the endogenous estrogen metabolism itself. Although all the used doses may not fall into a physiological concentration range, in cell culture models, isoflavone molecules were shown to increase the levels of enzymatic activity and/or of mRNA expression of the aromatase P450-*CYP19* (Edmunds *et al.*, 2005; Ye *et al.*, 2009) through the activation of the intra-cellular PKCα/P38/ERK1/2/CREB signaling pathways (Ye *et al.*, 2009). In contrast, isoflavones were also shown to down-regulate aromatase P450 and/or 3β-hydroxysteroid dehydrogenase (HSD) activities in other cell culture systems (Pelissero *et al.*, 1996; Blomquist *et al.*, 2005; Rice *et al.*, 2006; Wang *et al.*, 2008b). Besides all their powerful signalization properties (as shown later), another special characteristic of isoflavones is that it may have leverage with the steroidogenic enzyme activities in peripheral tissues; this might also be of interest in the putative neurosteroid-metabolizing enzyme-related dysfunctions during normal and/or physiopathological (neurodegenerative) aging of the hippocampal formation in human subjects.

Following the ingestion of a soy-derived product, the main isoflavones found in the circulating plasma are genistein and daidzein (Lephart *et al.*, 2004; Bennetau-Pelissero *et al.*, 2003; Vergne *et al.*, 2007). Equol has also been detected as circulating isoflavane in equine (Mariann and Haselwood, 1932), rodent (Lephart *et al.*, 2004, Mathey *et al.*, 2007) and ruminant (Woclawek-Potocka *et al.*, 2008) species. In some human subjects (approximately, between 40 to 60% of people consuming soy products, referred to as equol producers) equol has been detected in plasma and urine (Bennetau-Pelissero *et al.*, 2003, Mathey et al., 2006). Because of their lipophilic structure, circulating isoflavones in their aglycone form can cross freely all biological membranes including the blood-brain barrier. As a consequence, these molecules are able to reach all the physiological compartments and organs within the body, including the central nervous system, endocrine glands, cardiovascular system and squeleton (McEwen, 2002; Fitzpatrick, 2003; Wuttke *et al.*, 2003; Cornwell *et al.*, 2004; Ferrari,

2004; Lee *et al.*, 2005; Lephart *et al.*, 2005; Usui, 2006; Moutsatsou, 2007; Watson *et al.*, 2007; Xiao, 2008; Carlson *et al.*, 2008). These compartments and organs are vital sites for fine-tuned regulations of bodily homeostasis, of which regulations are deeply impacted by an intense signalization originating from imported (McEwen and Milner, 2007; Spencer *et al.*, 2008) and/or locally (Kretz *et al.*, 2004; Rune and Frotscher, 2005; Agis-Balboa *et al.*, 2006; Ish *et al.*, 2007; Hojo *et al.*, 2008) synthesized E_2. This occurrence was made possible because the upper mentioned compartments and organs are highly enriched in sites of specific and active interaction with E_2. Within the large class of E_2 effect-mediating molecules, new members of the three families of ERs (Maggiolini and Picard, 2010), ER-interacting proteins [PELP1, a praline-, glutamic acid-, and leucine-rich protein-1, (Brann *et al.*, 2008)] and non-ER/E_2-binding proteins (Valverde *et al.*, 1999; Curtis *et al.*, 2002; Sarkar *et al*,. 2008) are continuously still coming into light.

III.2. Brain Distribution
of Estrogen Receptors

In mammals, the brain is the conductor of most vital bodily functions that include metabolic and reproductive neuroendocrine regulations, as well as motor and cognitive behaviors. These brain functions are deeply impacted by E_2. Thus, documenting the neuroanatomical localization of ERs, precisely, is the first requisite to a broader deciphering of E_2 signalization throughout the brain. Therefore, the cerebral distribution of ERs has been documented using 6 major ensembles of morphological and molecular techniques:

1) Protein-binding probing. For example, radio-labeled agonist or antagonist,
2) Immunological detection of tissue protein antigens. For example, immunohistochemistry,
3) Immunological detection of electrophoretically-separated extracted protein antigens. For example, western blotting,
4) Complementary hybridization of tissue messenger ribonucleic acid (mRNA). For example, *in situ* hybridization,
5) Complementary hybridization of extracted mRNAs and subsequent electrophoretic separation of specific nucleotides. For example, ribonuclease protection assay,
6) Enzymatic *in vitro* amplification of extracted mRNAs. For example, reverse transcription-coupled polymerase chain reaction (RT/PCR) with qualitative, semi-quantitative, quantitative or real-time measurement variants.

It is noteworthy, that some alternative procedures could consist of a mixed protocol between 2 or more techniques from the aforementioned list of technologies. Nonetheless, the majority of resulting findings, but not all of them, have agreed to draw, somehow, a more or less coherent and unified picture between the relative patterns of distribution of ER-binding sites, ER mRNAs, and ER proteins. In the same context, the physical isolation of living cells arising from specific brain territories, along with the use of pharmacological tools, allowed neuroscientists around the world to gain invaluable insights into the physiological signature of E_2-mediated signalization.

III.2.1. Estrogen Receptor A

By means of Immunohistochemistry, the distribution of ERα protein was addressed in the rodent brain. The occurrence of ERα-immunoreactivity (ERα-IR) was shown to be clustered in brain areas such as the cerebral, prefrontal and piriform cortices, as well as the hippocampus, bed nucleus of the stria terminalis, medial nucleus of the preoptic area, medial nucleus of the amygdala, arcuate nucleus/periventricular zone of the hypothalamus, ventromedial hypothalamic nucleus, peri-acqueductal gray area, dorsal raphe nuclei, pons and locus coeruleus (Weiland *et al.*, 1997; Rune *et al.*, 2002; Mitra *et al.*, 2003; Rune and Frotscher, 2005; Kalita *et al.*, 2005; Mehra *et al.*, 2005; Ishunina *et al.*, 2007; Montague *et al.*, 2008; VanderHorst *et al.*, 2005; Hazell *et al.*, 2009, VanderHorst *et al.*, 2009). Within the hippocampal formation, the most heavily stained subdivisions were the cornu ammonis (CA) field 1-4 (CA1-4), dentate gyrus (DG) hilus and subiculum. The cerebral ERα immuno-staining was mainly visible in the nuclear compartment. In the hippocampus, ERα-IR was localized in cytoplasm, neuronal processes and nuclei as shown by both ERα-immunohistochemistry (Milner *et al.*, 2001; Rune *et al.*, 2002; Rune and Frotscher, 2005) and ERα-immunoblotting following cell fractionation (Kalita *et al.*, 2005).

In the hippocampal formation, the investigation of the mRNA expression confirmed the presence of ERα mRNA transcripts as shown by ribonuclease protection assay (Kalita *et al.*, 2005), by *in situ* hybridization (Rune *et al.*, 2002) and by RT/PCR (Ishunina *et al.*, 2007; Schreihofer and Redmond 2009).

III.2.2. Estrogen Receptor B

Immunohistochemical studies have documented the distribution of ERβ throughout the brain. The expression of ERβ-immunoreactivity (ERβ-IR) was shown to be less strong, but more extended throughout the brain, than the ERα-IR. Indeed, ERβ-IR was detectable in the cerebral cortex, piriform cortex, medial septum, vertical limb of diagonal band, bed nucleus of the stria terminalis, medial nucleus of the preoptic area, medial nucleus of the amygdala, arcuate nucleus/paraventricular nucleus/perventricular zone of the hypothalamus, supraoptic nucleus, ventromedial hypothalamic nucleus, hippocampus, reticular substantia nigra, ventral tegmental area, periacqudectal gray area, dorsal raphe nuclei, superior olive nucleus, locus coeruleus, spinal trigeminal nucleus and cerebellar Purkinje cells (Milner *et al.*, 2005; Nomura *et al.*, 2005; VanderHorst *et al.*, 2005; Hazell *et al.*, 2009). More specifically, within the hippocampal formation, the most evidently stained subdivisions were the CA1-3, DG, subiculum and entorhinal cortex (Rune *et al.*, 2002; Mitra *et al.*, 2003; Kalita *et al.*, 2005; Mehra *et al.*, 2005; Milner *et al.*, 2005; Rune and Frotscher, 2005; Chung *et al.*, 2007; Hazell *et al.*, 2009). More specifically, ultrastructural investigation of the ERβ immuno-staining revealed the presence of ERβ-IR in the plasma membrane, nuclear and cytoplasmic compartments, as well as in neuronal processes including dendrites and axon terminals (Milner *et al.*, 2005). These findings were confirmed by ERβ-immunoblotting following cell fractionation (Kalita *et al.*, 2005).

In the hippocampal formation, the investigation of the mRNA expression confirmed the presence of ERβ mRNA transcripts as shown by ribonuclease protection assay (Kalita *et al.*, 2005), by *in situ* hybridization (Rune *et al.*, 2002) and by RT-PCR (Kuiper *et al.*, 1997; Chung *et al.*, 2007; Schreihofer and Redmond, 2009).

III.2.3. G Protein-Coupled Estrogen-Receptor (GPR30)

The presence of GPR30-immunoreactivity (GPR30-IR) was demonstrated within all the major neuro-anatomical subdivisions of the brain including the piriform cortex, medial nucleus of the preoptic area, paraventricular nucleus, supraoptic nucleus, arcuate nucleus/peri-ventricular zone of the hypothalamus, striatum, medial nucleus of the amygdala, hippocampal formation, midbrain/pons area, locus coeruleus, spinal trigeminal nucleus and cerebellar Purkinje cells (Brailoiu *et al.*, 2007; Sakamoto *et al.*, 2007; Matsuda *et al.*, 2008; Hazell *et al.*, 2009). Within the hippocampal formation, the most heavily stained subdivisions were the DG hilus, subiculum, ventral part of the CA1-CA3 and entorhinal cortex. The GPR30 immuno-staining was mainly seen in cytoplasms and fibers in hippocampal neurons. The immunoblotting of GPR30 protein also confirmed its presence in the hippocampal formation of both male and female rats. The expression of GPR30 mRNAs was also detected in the CA1, CA2, CA3 and DG neurons (Matsuda *et al.*, 2008; Schreihofer and Redmond, 2009).

III.2.4. Conclusion

The documentation of $[^{125}I]$estrogen binding sites did corroborate some of the major findings obtained in immunohistochemical studies, specifically, in the hippocampus (Shughrue. and Merchenthaler, 2000; Milner et al., 2008). When comparing the relative distribution of each ER, we can observe that the co-existence of ER subtypes occur in many regions throughout the brain. Indeed, all the three ER subtypes were found to be expressed in numerous neuro-anatomical areas such as piriform cortex, medial nucleus of preoptic area, bed nucleus of stria terminalis, paraventricular nucleus, arcuate nucleus/periventriclar zone, ventrolateral part of the ventromedial hypothalamus nucleus, periacqueductal gray area, dorsal raphe nuclei, locus coeruleus, the spinal trigeminal nucleus and the hippocampal formation. Co-expression of both ERα and ERβ in the same hippocampal pyramidal cell layers, namely CA1 and CA3, was reported in both male and female rats (Rune et al., 2002). Furthermore, neuronal colocalization of both ERα and ERβ was also reported for the hippocampus (Mehra et al., 2005). The ultrastructural localization of ERα and ERβ revealed that these ERs could be found on the plasma membrane, as well as in nuclear, cytoplasmic and synaptic compartments of hippocampal neurons. The GPR30 protein was also linked to plasma membrane-localized functions. All together, these findings may imply that ERs play some important role in the mechanisms of rapid signalization in the hippocampal formation. Besides, ERs are, undeniably, robust transcription factors, regulating gene expression in countless estrogen-targeted tissues. When bound by E_2, GPR30 is also able to elicit a rapid production of cAMP (Thomas and Dong, 2006; Filardo et al., 2007) that may, ultimately, affect gene expression. Taken into account all the transduction features of E_2 signal through ERα, ERβ, and GPR30, it is conceivable that immediate, short-term and long-term signaling mechanisms may cooperate in concert in order to adequately shape the architecture of synapses animating the hippocampal neuronal networks. This would explain some of the well known and largely documented effects of E_2 on cognition and, more specifically, on hippocampal-dependent learning and memory functions. In addition, because ERα and ERβ can colocalize within the same hippocampal neurons (Mehra et al., 2005), it is reasonable to speculate that multiple ER-mediated signalizations do coexist in the same cells of the hippocampal formation.

Chapter IV

Brain Signalization by Estrogens and Estrogen Receptors

A large body of data made it evident that, beyond the reproductive function, ERs are both active players and molecular targets in a number of physiological regulations involving estrogen signalization in the mammalian higher central nervous system. Bioelectrical, neurochemical, structural, and behavioral aspects of E_2 signalization have been identified in the hippocampal formation of all mammalian species examined so far. In the present section, we are going to briefly describe some of the main aspects regarding estrogen signalization and the hippocampal formation physiology.

IV. 1. Estrogen Signalization through Estrogen Receptors

The mechanism of E_2-elicited genomic modulation of gene expression that has been extensively and repeatedly documented in various peripheral cell systems was also identified in the mammalian brain. Accordingly, at a number of brain sites that may include the hippocampal formation neurons, estrogens have been shown to regulate various genes such as low voltage-activated (T-type) Ca^{++} channel α_1 Cav3.1-isoform, tryptophan hydroxylase, serotoninergic 5HT1A/2A receptors, progesterone receptors, glutamic acid decarboxylase, anti-apoptotic bcl-2 proteins, BDNF, glial fibrillary acidic protein (GFAP), laminin, (McEwen, 2002; McEwen *et al.*, 1997; Qiu *et al.*, 2006; Scharfman and MacLusky, 2006; Foy *et al.*, 2008). Such a regulation certainly require cell trafficking of ERs from the cytoplasmic compartment to the nuclear location. Indeed, for example, E_2 induced nuclear translocation of ERα, but not of ERβ, to CA1 pyramidal neuronal nuclei (Rune *et al.*, 2002).

In the context of rapid signalization by E_2, ERα, and ERβ were localized in cell membranes where they initiate rapid effects through MAPK-dependent pathways (Zhao and Brinton, 2007a). Also, the expression levels of hippocampal GPR30 were shown to be similar in both genders in adult animals, but higher levels could be measured in male rat neonates (post-natal day 2, PND2) as compared to female rat neonates (Matsuda *et al.*, 2008). This suggests that E_2 signalization through GPR30 may play some dimorphic functional role in the developing hippocampus. In the female rat dorsal root ganglion (DRG), ovariectomy (OVX) resulted in decreased number of GPR30-IR neurons and estrogen replacement reversed this effect (Takanami *et al.*, 2010) implying that circulating ovarian steroids may regulate the transcriptional rate of the GPR30 gene. Elsewhere, in GPR30-transfected HEK293 cells (a cell model devoted of ERα, ERβ, and GPR30) E_2 was shown to bind isolated membrane preparation with high specificity and to elicit the production of cAMP within less than 20 minutes (Thomas and Dong, 2006). This confirmed the rapid-signal transducing function of membrane-associated GPR30 protein. In keeping with this, E_2-treatment of dissociated hippocampal cell cultures resulted in increased cAMP response element binding protein (CREB) phosphorylation (phosphorylated CREB, pCREB) within 1h, and this increased phosphorylaion was still detectable at 6h, 12h, and 24h post-treatment before it was down-regulated below the control level at 48h (S. Lee *et al.*, 2004). This activated intra-cellular signalization seemed to be triggered through both Ca^{++}-calmodulin-dependent kinase II (CaMKII) and mitogen-activated protein kinase (MAPK) pathways but, to a much lesser extent, through Akt pathway (S. Lee *et al.*, 2004).

IV. 2. Estrogen Signalization through Neuronal Bio-Electrical Activity

In neuronal cells, estrogen signalization can take place at the plasma membrane as a direct interaction with ion channel proteins or neurotransmitter receptors. Indeed, E_2 has been shown to bind directly to hippocampal L-type voltage-gated Ca^{++} channels (VGCC) and to potentiate VGCC-sustained currents (Sarkar *et al.*, 2008). Non-diffusible extra-cellular E_2, also, binds directly the large-conductance, voltage-gated, Ca^{++}-activated, potassium channel (MaxiK), regulatory β-subunit resulting in an increased sensitivity of the pore-forming, α-subunit (*Slo*), to intra-cellular Ca^{++} concentrations (Valverde *et al.*, 1999). In addition, the expression of *Slo* in the hippocampal formation was consistently demonstrated by *in situ* hybridization, immunohistochemistry, radioligand-binding, RT/PCR and western blotting, and the related protein was also localized to presynaptic elements including the mossy fibers and perforant path terminals (Knaus *et al.*, 1996, Pacheco *et al.*, 2008) implying that MaxiK channels may translate some E_2 signalizing effects to a rapid presynaptic neurotransmission at hippocampal principal neurons. In an epileptic model of rat (displaying high levels of recurrent spontaneous seizures) MaxiK signalization was dramatically down-regulated in the hippocampus (Pacheco *et al.*, 2008). Altogether, these findings suggest that MaxiK-mediated, E_2-elicited modulation of glutamatergic neurotransmission might play some significant role in the bio-electrical signalization of hippocampal formation neurons. Additionally, E_2 has been demonstrated to bind directly neuronal ACh receptor protein resulting in an allosteric change-mediated potentiation of the ACh-evoked currents under specific configurations of the α/β subunit-related composition of the channel (α_4/β_2 and α_4/β_4, but not α_3/β_2) (Curtis *et al.*, 2002). Respective mRNAs to these ACh receptor subunits have been detected in a large population of both human and rodent hippocampal neurons (Hellström-Lindahl *et al.*, 1999, Son *et al.* 2008), including glutamic acid decaboxylase-67 (GAD67)-expressing [most likely, γ-amino-butyric-acid (GABA)-synthesizing] inter-neurons (Son and Winzer-Serhan, 2008). In addition, E_2 modulation of bio-electrical activity involving GPCRs also occurs in brain neurons (Kelly and Wagner, 1999; Kelly and Levin, 2001; Kelly and Rønnekleiv, 2008). In a different context, E_2 did consolidate LTP induction through mitogen-activated protein kinase (MAPK)-dependent and -independent mechanisms in young (4-5 weeks-old) rats (Kim *et al.*, 2002).

IV. 3. Estrogen Signalization through Cerebral Neurotransmitter Systems

The brain distribution of ERs was shown to be closely related to serotoninergic, catecholaminergic and cholinergic neuronal systems (Nomura *et al.*, 2005; VanderHorst *et al.*, 2005; Aydin *et al.*, 2008; Yanagihara *et al.*, 2008; VanderHorst *et al.*, 2009) that, along side with GABAergic and glutamatergic neuronal systems, are well known to impact deeply the cognitive abilities (McEwen *et al.*, 1997; McEwen, 2002; McEwen and Milner, 2007; Maki and Dumas, 2009). These observations, and many others, have presently led to the widely spread notion for the existence of a robust estradiol-mediated control by the activity of brain neuronal neurotransmitter systems (Spencer *et al.*, 2008). Indeed, for example, Intra-cerebro-tissular administrations of E_2 into the cholinergic neuronal population-containing, medial septal-diagonal band of Broca area (Lâm and Leranth, 2003) or into the serotoninergic neuronal population-containing, median raphe nucleus (Prange-Kiel *et al.*, 2004; Prange-Kiel and Rune, 2006) result in a dramatic increase of spine density in the CA1 stratum radiatum. E_2-treatment also down-regulate hippocampal norepinephrine-elicited inositol phospholipid hydrolysis in OVX rats (Favit *et al.*, 1991) and serotonin-IR (Prange-Kiel and Rune, 2006). In the same context, administration of a high dose (by daily oral gavage with 1mg/kg b.w., for 6 weeks) of letrozole, an aromatase inhibitor, to normal intact rats resulted in increased levels of hippocampal monoamine (dopamine and noradrenaline) contents (Aydin *et al.*, 2008). Finally, local infusion of E_2 into the hippocampus did decrease the occurrence of serotoninergic fibers in both the stratum lacunosum moleculare and the stratum radiatum (Prange-Kiel and Rune, 2006) suggesting that E_2 increases the synaptic release and/or the enzymatic synthesis of serotonin at the raphe nuclei-arising serotoninergic axonal terminals.

IV. 4. Estrogen Signalization through Neuronal Morphological Adaptations

Morphological changes do occur in hippocampal formation neurons in response to E_2. Accordingly, E_2 is capable of shaping the architectural features of hippocampal afferent/efferent neuronal networks and both embed presynaptic and postsynaptic elements, alike. For example, E_2-treated OVX young rats displayed higher levels of axospinous synaptic density in the stratum radiatum as compared to vehicle-treated OVX rats of the same age (Adams *et al.*, 2001). Also, E_2 increases the spine density levels in the stratum radiatum molecular layer (Rune *et al.*, 2002). Additionally, local infusion of E_2 increases the number of spine synapses in the stratum radiatum of the CA1 region (Prange-Kiel and Rune, 2006). Finally, E_2 did up-regulate, whereas the ER antagonist, ICI 182,780, did down-regulate axonal outgrowth (VonSchassen *et al.*, 2006).

IV. 5. Estrogen Signalization through Synaptic Proteins

Biochemical adaptations do occur in hippocampal formation neurons in response to E_2. As a critical part of the E_2-mediated cellular adaptations, E_2 also modulates synaptic proteins, namely synaptophysin, spinophilin, post-synaptic density-95 protein (PSD95) and growth cone-associated protein-43 (GAP43). Indeed, E_2 increased the abundance of immunoreactive presynaptic protein, synaptophysin, in the stratum radiatum molecular layer (Rune *et al.*, 2002). Administration of E_2 to gonadectomized rats resulted in a dimorphic effect on stratum radiatum spinophilin-immunopositive fibers in the sense that E_2 increased spinophilin-IR in OVX female rats whereas it decreased spinophilin-IR in orchidectomized male rats, although E_2 did not affect hippocampal levels of spinophilin mRNA expression in both genders (S. Lee *et al.*, 2004). Forty-eight hours after E_2 treatment synaptophysin, syntaxin and spinophilin were up-regulated in the stratum radiatum and the stratum oriens of the CA1 region, and spinophilin also was up-regulated in the DG hilar area and in the stratum lucidum of the CA3 (Brake *et al.*, 2001). Similar results were obtained for the hippocampal CA1 region of the female Rhesus Macaques *(Macaca mulatta)*, after a menstrual cycle-long E_2-treatment, whereas no significant effect could be found in the molecular fields of the DG (Choi *et al.*, 2003). The modulatory effects of E_2 upon hippocampal synaptic plasticity could be ER subtype-specific in the sense that specific agonists to ERα or ERβ triggered different profiles of expression of synaptic molecules such as presynaptic proteins, synaptophysin, vesicular GABA transporter (VGaT), vesicular glutamate transporter-1 (VGlut1) or post-synaptic proteins, PSD95, spinophilin and AMPA-type glutamate receptors (AMPARs), GluR1/2/3 (Waters *et al.*, 2009).

IV. 6. Estrogen Signalization through Neurotrophins

In the context of the nurturing effects of a given brain structure toward its corresponding neuronal terminal inputs, the discovery of neurotrophins and their effects, did throw an invaluable light upon the intimate mechanisms of reciprocal signalization exchanges between the targeted brain area and the one that send axonal fiber afferents to this brain area. Hence, the hippocampal formation was identified as one of the major brain sites for the production of signalizing neurotrophins that are known to be deeply impacted by E_2. For example, it has been reported that E_2 increases hippocampal NGF mRNA levels (Pan *et al.*, 1999) and the resulting protein increases TrkA and ERK/MAPK activity within the hippocampus-projecting medial septo-basal forebrain neurons (Williams *et al.*, 2006). These hippocampal physio-molecular modifications in response to E_2 administration were associated with improved spatial working memory (Pan *et al.*, 2000) through a better recruitment of cholinergic neurons (Pan *et al.*, 1999). In the same context of E_2-triggered production of neurotrophins, the hippocampus was shown to produce brain-derived neurotrophic factor (BDNF) which, in turn, affects local synthesis of neuro-peptide Y (NPY) (Scharfman and MacLusky, 2006). Also, BDNF-mediated E_2-elicited signalizations exert, locally, deep impacts on neurotransmitter neuronal systems arising from the hippocampal formation or from other brain areas.

IV. 7. Estrogen Signalization through Hippocampal Estrogens

Another interesting aspect of E_2 signalization resides in the fact that this molecule can also be considered as a neurosteroid. Indeed, E_2 is synthesized locally within neuronal and glial cells of the hippocampal formation (Hojo *et al.*, 2008; Kretz *et al.*, 2004; Rune and Fortscher, 2005; Agis-Balboa *et al.*, 2006; Ish *et al.*, 2007; Garcia-Segura, 2008) where the testosterone-aromatazing enzyme, aromatase (cytochrome P450 depending enzyme *CYP19α*), and the dehydro-epiandrosterone synthase (cytochrome P450 depending enzyme *CYP17α*), were detected preferentially in pre- and post-synaptic ultrastructural elements of the CA1-3/DG principal neurons (Hojo *et al.,* 2004). In the Rhesus monkey hippocampus, aromatase-IR is present in CA1-3/DG within NeuN (a neuronal marker)-, but not GFAP-coexpressing cells, implying that aromatase is preferentially expressed in neurons (Yague *et al.,* 2008). In addition, the hippocampal aromatase-IR, that was highly enriched in dendritic processes suggesting that the aromatase enzyme may play some key role in the mechanisms of synaptic plasticity, did also display interesting co-expression profiles within hippocampal calretinin-, calbindin- and parvalbumin-immunopositive neurons (Yague *et al.,* 2008). The pharmacological blockade of hippocampal endogenous E_2 synthesis using the aromatase inhibitor, letrozole, or an RNA silencer (siRNA) directed against steroidogenic acute regulatory (StAR) protein, resulted in a significant down-regulation of spine formation as well as of synaptic proteins, namely synaptophysin, spinophilin and GAP43 (Rune *et al.,* 2002; Rune and Frotscher, 2005). In keeping with this, letrozole administration to normal intact rats resulted in increased expression levels of hippocampal neural cell-adhesion molecules (N-CAM), the 140 and 180 isoforms (Aydin *et al.,* 2008). Aromatase knockout (ArKO) female mice, that are estrogen deficient animals, displayed significantly higher mRNA expression levels of hippocampal NMDA receptor subunit NR2B, and to a lesser extent, that of NR1 and NR2A, as compared to wild-type animals (Boon *et al.,* 2005). Treatment of hippocampal slices with letrozole, resulted in a down-regulation of spinophilin-IR and both densities of spine synapses and buttons in the stratum radiatum (Prange-Kiel and Rune 2006). Letrozole administration also resulted in a dose-dependent, down-regulation of hippocampal neuronal synaptophysin-IR in dispersed cell cultures (Prange-Kiel *et al.,* 2006). Finally, letrozole and the ER antagonist, ICI182780, did down-regulate the expression levels of GAP43 (VonSchassen *et al.,* 2006).

IV. 8. Estrogen Signalization through Estrogen Receptor-Interacting Proteins

It is noteworthy to evoke some other specific mechanisms of ER-mediated signalization occurring in the hippocampal formation and involving a different type of signaling molecules, such as PELP1. This protein is a steroid receptor-interacting molecule which is implicated in the transcriptional regulation of gene expression in steroid targeted cells. Indeed, PELP1 was shown: (1) to interact with ERα, probably with ERβ, too; (2) to be expressed in a number of

E_2-targeted, ER-expressing brain regions including the hippocampus; (3) to be co-expressed within hippocampal ERα-immunopositive neurons; (4) to be recruited to the promoters of ERα-targeted genes such as progesterone receptor, pS2 and insulin-like growth factor; (5) to modulate cyclin D1 mRNA expression; and (6) to be associated with an increased phosphorylation of STAT3 (at least, when PELP1 is over-expressed in MCF7 cells, and given the fact that estrogen enhances brain phosphorylation of STAT3) through Src-MAPK-dependent pathway (for an extensive review on PELP1, see Brann *et al.*, 2008).

IV. 9. Estrogen Signalization Modulating Mnemonic Behaviors

In mammals, their animal behaviors are deeply impacted by estrogen signalizations. Specifically, the mnemonic characteristics constitute, most likely, a behaviorally expressed translation of a large part of the biological effects we reported earlier are triggered, directly or indirectly, by estrogens at the cognitively competent brain sites. For example, E_2-replacement in OVX rats improved the performance of spatial working memory monitored in an 8-arms radial maze when E_2-treatment was administered soon after the surgical removal of the ovaries but not after 5 months of E_2 withdrawal (Daniel *et al.*, 2006). Similar results were obtained in the Morris water maze (Talboom *et al.*, 2008). This implies that circulating estrogens are essential for the maintenance of efficient functioning in the cognitively competent brain areas at least in females. It does also suggest that the ability of the female hippocampus to synthesize its own E_2 is not sufficient for a normal behavioral expression of mnemonic acquisition, or that this ability is impaired under long-term E_2 withdrawal. Moreover, peripheral administration (0.2mg/kg b.w.) or intra-tissular dorso-hippocampal infusion (5.0μg/side) of E_2 resulted in enhanced object recognition memory consolidation through a membrane-initiated ERK activation (Fernandez *et al.*, 2008). Contrasting results were reported throughout the oestrous cycle where increased levels of plasma E_2 were associated with diminished performance of spatial memory as assessed in the Morris water maze (Frye, 1995). This occurrence led to the suggestion that E_2-mediated increase of newly formed synapses in the hippocampal area could be detrimental to the spatial reference memory. However, agonist-specific activation of ERβ, but not of ERα, was associated with increased expression of hippocampal synaptic proteins, increased LTP potentiation and improved performance in nonmatch-to-sample tasks in hippocampus-dependent spatial memory (Liu *et al.*, 2008). Quite surprisingly, though, letrozole administration to normal rats, also, did improve the consolidation of spatial memory (Aydin *et al.*, 2008). Therefore, the question of the dose and/or of the activated ER-subtype, seems to be essential. However, ArKO female mice, did acquire the spatial reference memory task in the Morris water maze as well as wilde-type mice did (Boon *et al.*, 2005). Altogether, these findings show that there is no simple equation between the E_2 level and the cognitive memory performance. The improvement of memory under low levels of E_2, might suggest that neurophysiological adaptations are still taking place in the E_2-deprived brain in order to rescue hippocampus-dependent spatial memory functions. Discovering some key mechanisms within these neurophysiological adaptations may benefit greatly for the rescue of the aging hippocampus and, more particularly, in the neurodegenerating and/or postmenopausal brains.

Chapter V

Natures of Aging

Nature has sentenced every single matter that has succeeded in even a slightest occupation of time, to an irrevocable aging process. Accordingly, both mineral and organic matters do experience more or less severe deteriorations with increasing age. In living organisms, these deteriorations inevitably concern the species as a whole entity, the population as an ensemble of individuals and the living individual subject throughout different stages of its lifespan, as well as both the physiological function within every single organism and the behavioral expression of each single living thing.

V.1. Aging Species

Phylogenetic evolutions of living organisms within both the animal and the vegetable kingdoms have taught us that species do appear in a given stage during earth's ages. Species also develop and strengthen over time. Sometimes they do successfully thrive and multiply to constitute large populations. Evolutionary successful species usually expand their occupation of space beyond their original inhabited ecosystem. This occurrence often throws the emerging species into the arena in the struggle for survival. After a more or less longer period of time, the formerly flourishing species do decline and then after disappear for multiple reasons. These reasons might be attributable to suddenly or gradually adverse environmental changes (physical or biological changes) and/or to an inherent age-related accumulation of crippling biological defects. The later reason could be due to different natures such as a genomic nature (life expectancy-reducing, immunogenic deficiency-inducing or reproductive age-shortening gene mutations) or a behavioral nature (excessive food-related specialization, for example). These phenomena and many plausible others can play a central role in the aging processes of a given species on the pace toward the dawn of extinction.

V.2. Aging Individuals and Aging Populations (Humans)

In different parts of the world, human populations are increasingly growing and aging (Christensen *et al.*, 2009; Lutz and Qiang, 2002; Shrestha 2000). For example, an estimated 420 million persons were aged ≥ 65 years in the year 2000 (CDC, E003). This number was the result of a 9.5 million increase that was recorded within one year time laps (CDC, E003). The worldwide number of individuals aged ≥ 65 years is projected to increase by approximately 550 million to 973 million during the period 2000-2030 (CDC, E003). These changes of demographic features are likely due, at least in the most industrialized countries, to decreased fertility and increased life expectancy that have been obviously seen in a number of human societies worldwide during the 20th century. Consequently, the ranking within the list of leading causes of deaths is changing in favor of neurodegenerative diseases, such as Alzheimer's disease (AD) (Ferrucci *et al.*, 2008). Old age-related disability, morbidity and co-morbidity are also major concerns for both public health service planners and providers alike, around the world (Ferrucci *et al.*, 2008). Hence, the aging-triggered physiological dysfunctions result in an increasingly worsening social burden upon worldwide human societies. Because nutrition is able to deeply impact the functioning of bodily organs, and also by improved hygiene and pharmaceutical/medical technologies, are the leading causes of increased life expectancy. Ironically, though, nutrition remains the ultimate and highly needed gateway we are hoping for. In order to partially alleviate age-associated physiological dysfunctions where the memory decline may be bringing along the most costly and challenging social burden that humanity must face as the present century begins. However, carefully examined and well documented dietary/supplemental isoflavone-related recommendations are still highly missing.

V.3. Aging Function (Mammalian Reproductive Neuroendocrine Axis)

In the mammalian brain, gonadotrophin-releasing hormone (GnRH) neurons, which lay in the preoptic/medio-basal hypothalamic areas, are well known to orchestrate the activity of the hypothalamo-pituitary-gonadal axis. By providing the endocrine pituitary cells with phasic pulses of the signalizing GnRH peptide, these neurons are able to entrain the pituitary secretion of both luteinizing hormone (LH) and follicle-stimulating hormone (FSH) in an oscillatory fashion. Down-stream of the blood circulation, LH and FSH are addressed to the ovaries which in turn, as a response to the pituitary hormonal signal, synthesize and release estrogens in a cyclic manner. Back into the brain, the later conveys a retro-control message to GnRH neurons directly or via numerous mono-aminergic/peptidergic neuronal systems. Other molecules emanating from the ovary, such as for example inhibin A/B or gonadotrophin surge-attenuating factor (GnSAF), do participate in the establishment of positive/negative sensing retro-controls exerted by the ovary upon the pituitary FSH/LH secretion. The major players within this functional loop are grouped under the name of reproductive neuroendocrine axis/system (for reviews on this axis, see Smith and Jennes, 2001; McEwen, 2002; Messinis, 2006; Spencer et al., 2008). Therefore, the reproductive neuro-endocrine system is responsible for cyclic features in the basal free-running physiology of most, if not all, the organs within the body. Among the organs affected by the fluctuating E_2 plasma levels, the brain is one of the most sensitive organs to circulating E_2. Within the brain, the hippocampal formation is a site of active integration of the ovary-emanating hormonal signals. Consequently, overall neuronal plasticity adaptations, namely bio-electrical, physio-molecular, neurochemical and morphogenic adaptations do occur throughout the oestrous cycle in the hippocampal formation. For example, in the mouse hippocampus, the activity markers of Akt, LIM kinase, TrkB, synaptophysin and PSD95 all undergo oestrous cycle-shaped fluctuations, although the performance in an object placement task remained unaffected by the oscillating endogenous estrogens (Spencer et al., 2008). However, the reproductive neuro-endocrine system is subjected to age-over time alteration processes resulting in major accumulating dysfunctions (Downs and Urbanski, 2006; Hall, 2007; Downs and Wise, 2009). Three major aspects in the functioning of the aged reproductive neuro-endocrine axis are the loss of hypothalamic GnRH pulse secretion, the dramatic raise in LH/FSH secretion and the overwhelming down-regulation of ovarian/testicular estrogen/androgen synthesis/secretion (Downs and Urbanski, 2006; Messinis, 2006; Hall, 2007; Downs and Wise, 2009).

V.4. Aging Cognition (Mammalian Hippocampal Formation Sub-Serving Memory Function)

Similarly to other vital organs within the body, the brain is deeply impacted by the passage of time (Crews, 2007). As a result, the aging of overt cognitive abilities can be markedly and increasingly seen in age-advanced populations. More specifically, the hippocampal formation is no exception in the age-impacted physiologically altered cerebral structures. Thereafter, age-related hippocampus-dependent learning and memory dysfunctions do occur in elderly people. This aspect of cognitive aging has been intensively studied and its highly multifaceted and complex feature has been documented and extensively reviewed elsewhere (Halbreich *et al.,* 1995; Bellino and Wise, 2003; Billard, 2006; Dumas *et al.,* 2008; Barrett-Connor and Laughlin, 2009; Frick 2009). Because the cell firing of hippocampal neurons was correlated to the position of the animal in its environment, this discovery and many others allowed the hatching of the ever alive concept of the hippocampus being the brain site where the spatial map is built-up using the integrated surrounding visual cues (Hill, 1978; Olton *et al.,* 1978; McNaughton *et al.,* 1983). This functional characteristic of the hippocampus shows it as the mastermind of a highly developed spatial navigation art in mammals. Unfortunately, bio-electrical, structural and neurochemical alterations, more or less severely, do occur in the aging hippocampi of rodents, non-human primates and humans and these alterations are believed to sustain behaviorally expressed, hippocampus-related cognitive defects. Indeed, aging has been associated with altered inductions of both long-term potentiation (LTP) and long-term depression (LTD) in hippocampal neurons and with hippocampus-dependent learning deficits that was shown in rodents, non-human primates and humans (Foster, 1999; Foy *et al.,* 2008; Frick, 2009; Lacreuse, 2006; Genazzani *et al.,* 2007). In addition, induced LTP in CA1 pyramidal neurons arise from the fusion of two different conductances, an NMDA receptor (NMDAR)-dependent component and a voltage-dependent calcium channel (VDCC)-related component (Shankar *et al.,* 1998). Quite interestingly, and compared to young rats (6-9 week-old), aging rats (24 month-old) displayed a reduction in the magnitude of the NMDAR-dependent component and an increase in the magnitude of the VDCC-dependent component (Shankar *et al.*, 1998) implying that compensatory mechanisms against the age-triggered alterations of neuronal excitability, are shifting the bio-electrical activity of hippocampal neurons to much higher Ca^{++} concentrations in old subjects as compared to young ones.

Age-related alterations in the hippocampal formation physiology have also been associated with impaired E_2 signalization and, more particularly, ER-related signalization (for an extensive review, see Frick, 2009); and, thereof, aging women health is more at stake (Sherwin, 2006; 2007). For example, the number of both ERα and ERβ immunopositive neurons was shown to be affected in the hippocampus during the normal aging process. Indeed, the number of ERα immunoreactive neurons decreases by 78% and 56% in, respectively, the CA3 and the CA1 of aged female rats as compared to normal adult female rats (Mehra *et al.*, 2005). Also, the number of ERβ immunoreactive neurons drops by 88% and 41% in, respectively, the CA3 and the CA1 of aged female rats as compared to normal adult female rats (Mehra *et al.*, 2005). In the same study, western blotting did confirm that the amounts of ERα and ERβ proteins were highly decreased in aged female hippocampi. In the same context, ERα-IR and aromatase (P450arom)-IR were up-regulated in post-menopausal women hippocampi whereas ERα and P450arom mRNA transcripts were down-regulated in Alzheimer disease (AD)-deceased women hippocampi (Ishunina *et al.*, 2007).

In the context of age-triggered alterations of human hippocampal nicotinic ACh receptor (nAChR)-mediated neurotransmission, it was shown by mean of RT/PCR that the mRNA expression levels of the β_2 subunit, but not the α_4 subunit, is down-regulated in post-mortem hippocampi of elderly people (Tohgi *et al.*, 1998). Similarly, the mRNA expression levels of the α_7 subunit, but not the α_3 or α_4 subunits, were up-regulated, whereas the $[^{125}I]$-α-bungarotoxin binding levels were down-regulated in post-mortem hippocampi of Alzheimer disease patients (Hellström -Lindahl *et al.*, 1999). Furthermore, an altered signalization occurs, in aged rats, in the cholinergic basal forebrain-hippocampal pathway as revealed by a down regulation of hippocampal nerve growth factor-induced TrkA expression and ERK phosphorylation (Williams *et al.* 2006) which is believed to be crucial for memory encoding (Giovannini *et al.*, 2005; Giovannini 2006). Recently estradiol has also been shown to interact at the hippocampal level with the cholinergic receptors as a protective agent of the cerebrovascular system that is altered during aging (Kitamura *et al.*, 2009). In human hippocampi, structural and neurochemical changes have been demonstrated by means of magnetic resonance imaging and proton magnetic resonance spectroscopy in two hippocampus-dependent memory tasks, namely virtual Morris water maze and transverse patterning discrimination tasks (Driscoll *et al.*, 2003).

Under the paradigm of hormone replacement therapy, in both aged female and aged male mice (24 months of age), E_2-treatment improved memory performance in inhibitory avoidance and water maze tasks (Frye *et al.*, 2005). E_2 benzoate (EB)-treatment of aged mice (27-28 month old) also improved spatial reference memory performance in high EB-dose (5 µg/day) injected but not in low EB-dose (1 µg/day) injected animals (Frick *et al.*, 2002). Aged mice treated daily with 5 µg EB displayed increased expression levels of hippocampal synaptophysin-immunoreactivity, whereas the activity levels of hippocampal ChAT and GAD remained unaffected by the treatment (Frick *et al.*, 2002). Elsewhere, E_2-treatment did increase axospinous synaptic density in the stratum radiatum of OVX young ($\approx$ 4 months) rats but not in that of OVX aged ($\approx$ 24 months) rats (Adams *et al.*, 2001). On the contrary, E_2-treatment did increase the number of NMDA receptor, NR1 subunit, in individual synapses of the CA1 stratum radiatum of OVX aged rats but not that of OVX young rats (Adams *et al.*, 2001) suggesting that some lower energy-requiring cellular adaptations are taking place in the aging hippocampal neurons as an ultimate attempt to counteract the increasing age-linked defects of hippocampal neuronal excitability and/or of E_2 signal transduction.

Chapter VI

Pharmacological Effects of Isoflavones on the Hippocampal Formation

In order to decipher the impact of isoflavones on the memory function, we did collect two main ensembles of data that differ by their delivery routes of active molecules. The first ensemble of data consisted of using isoflavone molecules as pharmacological tools. This was mainly carried out in simplified hippocampus-derived cell systems including organotypic hippocampal slice cultures, dissociated hippocampal cell cultures (neuronal + glial cells) and hippocampus-derived synaptosomal membrane preparations. The corresponding data are summarized in Tables 1.1-6, where the studies are classified in the decreasing order of the isoflavone concentrations (from, nutritionally, the less plausible to the highly probable). The second ensemble of data consisted of using isoflavone molecules as dietary compounds (for example, those naturally isoflavone-rich products) or as orally-administered, supplemental compounds (for example, those with a molecularly controlled composition). The corresponding data are summarized in Tables 2.1-3, for studies in rodents; in Table 3, for a study in a non-human primate (Rhesus monkey); in Tables 4.1-5, for a sample of cognitive testing in human subjects; in Tables 5.1-8 for dietary/supplemental interventions in human subjects.

Pharmacological effects of isoflavones on hippocampal neurons have been examined in different mammalian species including humans (control and AD-deceased subjects), rats (Wistar and Sprague-Dawley strains), mouse and guinea pig (see Table 1. for a complete summary). For most of them, these studies consisted of *in vitro* investigations that used hippocampus-derived cell systems arising from embryos, neonates, young or adult subjects. However, few studies have explored the effects of isoflavones in whole animals. The mostly investigated isoflavone molecule, by far, has been genistein. This is most likely due to its inherent ability to act as a powerful competitor at the ATP-binding site on protein tyrosine kinases (PTKs), and thereof, as a potent inhibitor of intracellular PTK activity (Levitzki, 2006). Besides, PTKs have been involved in a broad range of intra-cellular signalizations taking place ubiquitously in different neuronal systems, not to mention non-neuronal ones, which allow genistein to affect count-less intracellular signalization pathways. Within the scale of data availability, daidzein ranks as the second mostly studied isoflavone molecule in hippocampus-derived cell systems. Much fewer data did report effects of other isoflavones, namely formononetin or of the isoflavone metabolite, the isoflavane, equol (Setchell *et al.*, 2002).

VI.1. Description of Pharmacological Studies

Pharmacological studies carried out using hippocampal neurons did address isoflavone signalization under four main paradigms that correlate to four main modulations of the neuronal activity:

(1) bio-electrical modulation, consisting of measurable changes in the pattern of electrical activity of electro-physiologically recorded hippocampal neurons,
(2) physio-molecular modulation, consisting of quantitative and qualitative changes in the molecular phenotype of hippocampal neurons in response to a given signaling substance,
(3) neuroprotective modulation, consisting of measurable pro-survival and anti-apoptotic impacts on a given hippocampal neuronal population that has been challenged by a cell stressor,
(4) neurogenic modulation, consisting of visually detectable transformations, under microscope, of the cellular structure of hippocampal neurons that have been exposed to a given neurogenesis inducing molecule.

These four features of isoflavone signalization have been widely documented for estrogen signaling molecules in which E_2 has lead as the reference molecule. In the pharmacological studies, as summerized later, genistein was mainly used as a specific tyrosine kinase inhibitor and daidzein was sometimes used in parallel as a biologically inactive structural analog of genistein with no tyrosine kinase inhibitory effect.

VI.1.1. Bio-Electrical Features

The Isoflavone effects on bio-electrical properties of hippocampal neurons have been addressed by mean of electrophysiological recordings in a number of cell configurations. O'Dell et al. (1991) used guinea pig hippocampal slices to investigate the effect of genistein (110 μM) on the Schaffer collateral/commissural fiber stimulation-induced long-term potentiation (LTP) in the CA1 pyramidal neurons. The results showed that, when applied before the titanic stimulation, genistein was able to block both the induction of LTP and the activation of a hippocamus-enriched PTK (namely, the pp60c-src[+]: a cellular counterpart of Rous sarcoma virus tyrosine kinase) in CA1 pyramidal neurons. Similarly, Shankar et al. (1998) have used hippocampal slice preparations taken from young (aged 6-9 weeks) or old (aged 24 months) male, virgin Fisher 344 rats, to stimulate the Schaffer collateral/commissural fibers and to examine the effect of pharmacological genistein on electrical stimulation-induced LTP in the CA1 pyramidal neurons. They have demonstrated that the normal aging process was associated with a reduced magnitude of the NMDAR-dependent component and an increased magnitude of the VDCC-dependent component as of the relative contributions of these current components to the LTP induction (Shankar et al., 1998). Quite nicely, these authors have also demonstrated that 20μM genistein reduces LTP induction in young rats and almost block it completely in aged rats. In the same context, Casey et al., (2002) have demonstrated that intra-cerebro-ventricular injection of 250 μM genistein prior to the electrical stimulation of the perforant path leads to the abolition of the expected induction of LTP in the DG granule cells. This abolition was also associated with a complete blockade of Glutamate release and Ca^{++} influx normally observed in potassium chloride (KCl)-stimulated hippocampal synaptosomes, subsequently prepared from electrically-stimulated rat brains.

Ramsey et al., (2005) used hippocampal slices derived from young Sprague-Dawley rats to demonstrate that acute application of the insulin-like growth factor-1 (IGF-1) results in a dramatic increase of an α-amino-3-hydroxyl-5-methyl-4-isoxazolepropionate receptor (AMPAR)-mediated excitatory synaptic transmission in the CA1 region, and that previous bathing of hippocampal slices with genistein (220 μM) completely blocks this increase. Genistein (100 μM), but not daidzein (100 μM), was also able to alleviate a metabotropic glutamate receptor (mGluR)-mediated inhibition of Ca^{++}-activated potassium (K^+) currents [also called after-hyper-polarization K^+ currents (I_{AHP})] in DG neurons recorded on Wistar adult male rat-derived hippocampal slices (Abdul-Ghani et al., 1996). Similarly, genistein(50 μM) , but not daidzein (50 μM), was able to block 4β-phorbol ester myristate acetate (4β-PMA)-mediated potentiation of N-Methyl-D-Aspartate (NMDA)-activated currents in isolated hippocampal CA1 pyramidal neurons (Lu et al., 1999). In keeping with this, genistein (100 μM) but not daidzein (100 μM), was shown to reduce a depolarization-coupled induction of high voltage-activated Ca^{++} currents in CA1 pyramidal neurons (Potier et al., 1999). Genistein (100 and 250 μM) was capable of potentiating ACh-evoked currents in CA1 interneurons which were, presumably, GABA-ergic neurons (Charpantier et al., 2005). The same result was obtained by Cho et al., (2005) with Genistein (10 μM). However, daidzein (10 μM) was unable to potentiate the ACh-evoked currents in the same neurons (Cho et al.. 2005).

VI.1.2. Physio-Molecular Features

Due to their crucial role in the neuronal signalization, the activity of ion channels was documented in the presence of isoflavone molecules in a number of hippocampus-derived cell models. This time again, genistein was mainly used as potent tyrosine kinase inhibitor. Considering the hormonal regulation of ion channel availability on the cell membrane surface, genistein (1 μM) did block efficiently the insulin-elicited, but not the AMPA- or NMDA-elicited, internalization of AMPARs seen in dissociated DG neurons taken from PND0 neonate rat pups (Beattie et al., 2000). The effects of isoflavones were also examined under the paradigm of neurotransmitter-mediated regulations of ion channel trafficking. Charpantier et al., (2005) investigated the ACh-induced internalization of ACh receptors (AChRs) in cultured hippocampal slice neurons from 2-5 weeks aged rats. They demonstrated that genistein (100 and 250 μM) was: (1) capable of blocking the tyrosine phophorylation of the ACh nicotinic receptor α-7 (α7nAChR) subunit; but, (2) not capable of inducing the translocation of this subunit to the plasma membrane. Similarly, Mielke and

Mealing (2009) have shown that the maximal pool of α7nAChRs was localized in the intra-cellular compartment and that genistein (100 μM), insulin and K^+-triggered depolarization were all unable to elicit a significant translocation of the intra-cellular α7nAChR subunits. On the contrary, Cho *et al.* (2005) have reported a significant increase in the number of cell surface-localized α7nAChR subunits under the conditions of genistein (10 μM) treatment. Additionally, site-directed mutation of each, or all of the 3 cytoplasmic tyrosine residues, did not prevent the genistein-mediated facilitation of ACh-evoked currents in transfected Xenopus oocytes implying that a direct phosphorylation/dephosphorylation mechanism is unlikely. Elsewhere, in hippocampal membrane synaptosomes prepared from AD-deceased or control human subjects, genistein dose-dependently (from 1 to 10 μM) stimulated the binding of sulphate 35-radio-labeled GTP-gamma stimulating subunit ($[^{35}S]$-GTPγ-S) to synaptosomal preparations. This effect of genistein (1 to 10 μM) was similar in both AD and control hippocampi and could also be mimicked by daidzein from 1 to 10 μM (Jefremov *et al.*, 2008) indicating a non-genomic estrogenic effect rather than an anti-tyrosine kinase effect.

VI.1.3. Neuroprotective Features

Estrogens are well known for their robust actions as molecular shields against cell damages, and more particularly against neuronal damages that are caused by cell stressors. For this, they have, deservedly, been labeled as neuroprotective molecules. Indeed, the neuroprotective features of estrogens have been largely reviewed in previous well-documented literature (Garcia-Segura *et al.*, 2001; Wise *et al.*, 2001; Brann *et al.*, 2007). Because isoflavones could mimic different physiological regulations normally devoted to E_2, the putative capacity of these molecules to protect brain neurons, have been challenged in experimental conditions of cell stress.

VI.1.3.A. In Vitro *Approaches*
The study of the effect of genistein on intra-endoplasmic reticulum Ca^{++} stores was undertaken *in vitro* using thapsigargin which is believed to inhibit the sarco/endoplasmic Ca^{++} ATPase (SERCA) pump. This results in a dramatic depletion of intra-endoplasmic reticulum Ca^{++} stores and leads, subsequently, to the activation of store-operated Ca^{++} channels (SOCCs) which are known to be responsible of massive Ca^{++} influx into the

thapsigargin-activated cell. Using thapsigargin along with the Fura-2 Ca^{++} imaging, Koss et al. (2009) were able to show that genistein (100 or 10 µM depending on the time of treatment) remarkably reduces both the amplitude and the latency of the peak response of SOCC-mediated thapsigargin-elicited Ca^{++} influx into dissociated hippocampal neurons isolated from Sprague-Dawley PND3-5 rat pups.

In the configuration of neuroprotection against chemical insult, primary hippocampal neuronal cultures derived from PND1 rat pups were used to investigate the β-amyloid-induced neurotoxicity. The results, interestingly, highlighted the anti-apoptotic mechanisms of the isoflavones within hippocampal neurons. Indeed, Zeng et al., (2004) have shown that, in the β-amyloid-induced neurotoxicity model, genistein (40 µM) was efficiently able: (1) to decrease both the resulting DNA fragmentation and reactive oxygen species (ROS) production; and, (2) to block the associated increases in intra-cellular Ca^{++} concentrations ($[Ca^{++}]_i$) and pro-apoptotic caspase-3 activity. This study also demonstrated that the anti-apoptotic neuroprotective effects of genistein were completely abrogated by addition to the neuronal culture medium of the ER antagonist, ICI 182,780; to support that the isoflavone may act through an ER-binding, gene expression-regulating mechanism to protect hippocampal neurons against the β-amyloid neurotoxic insult. Elsewhere, ED18-derived primary neuronal cultures were used to investigate the neuroprotective effects of genistein in glutamate- and β-amyloid-induced neurotoxicity models. The results showed that, in both models and compared to control, low concentrations of genistein (<10 µg/mL) were efficiently able to rescue the plasma membrane integrity (as evidenced by the decreased lactate dehydrogenase (LDH) depletion index) but not the mitochondrial metabolism (as evidenced by the unaltered levels of 3-[4,5-di-Methyl-Thiazol-2-yl]-2,5-diphenyl Tetrazolium bromide (MTT) reduction index) (Zhao et al., 2002). In this study, daidzein, formononetin, and equol all have been reported to have similar neuroprotective effects to genistein; demonstrating, though, an estrogenic effect rather than an anti-tyrosine kinase effect. However, up to 100 ng/mL (i.e., 0.37 µM) concentration, genistein did not display any noticeable neurotrophic effect upon ED18-derived cultured hippocampal neurons (0 parameter affected, out of 6 morphological parameters that were visibly enhanced by 10 ng/mL E_2). In addition, at 10 µg/mL (i.e., 37 µM) concentration, genistein, but not daidzein, was rather neurotoxic for the studied neurons (Zhao et al., 2002) probably because of anti-protein kinase effects. On the contrary, glutamate-mediated neuronal apoptosis of dissociated hippocampal neurons isolated from ED15-17 Swiss mouse embryos was efficiently and dose-dependently opposed by increasing concentrations of genistein (0.01 to 10 µM) in the culture medium (Kajta *et al.* 2007). Again, the effects of genistein were completely abrogated by the ER antagonist, ICI 182,780. The effect is stronger in hippocampal neurons than in the other types of neurons tested, and genistein already exerts a protective effect at 0.01 µM. This low dose elicited effect would advocate for an estrogenic membrane effect. Inversely, these effects were greatly potentiated by a glycogen synthase kinase-3β (GSK-3β) inhibitor. Finally, the authors of this study have also shown that, up to a 10 µM concentration, genistein did not trigger any neurotoxic effect in hippocampal neurons (Kajta *et al.*, 2007).

VI.1.3.B. In Vivo *Approaches*

Azocoitia and colleagues (2006) studied the neuroprotective effects of a single intra-peritoneal (i.p.) injection of a soy extract preparation (containing 40% of isoflavones, 0.2, 1, 2, 20mg/kg body weight, b.w.) or increasing concentrations of genistein (0.1, 1, 10 mg/kg b.w.) upon the kainic acid (KA)-induced degeneration of hilar DG neurons in 7-day-ovariectomized (OVX) Wistar albino female rats. Both of the highly dosed injections of soy extracts (1, 2, 20 mg/kg b.w.) or genistein (10 mg/kg b.w.) were efficiently proactive in counteracting the neuronal loss within the hilus area of the DG. Interestingly, when injected alone at the highest dose, each one of the tested isoflavone preparations could not initiate any evident adverse effect (namely, neuronal loss) in the hippocampal formation. Xu *et al.*, (2008) examined the effects of genistein, given as a short-term treatment (7-12 days, 250 µg/day) to OVX Sprague-Dawley female rats, upon the OVX-induced impairment of hippocampal function. Two parameters were addressed in this study: the ultrastructural feature of mitochondria and the ATP contents in the hippocampal CA1 region. It was found that genistein treatment opposed both OVX-induced impairments of hippocampal CA1 neurons, namely morphological abnormalities of mitochondria (but not their numerical density) and decreased ATP contents. As of a different model of neuronal stress, thapsigargin is believed to inhibit the sarco/endoplasmic Ca^{++} ATPase (SERCA) pump and to result in a dramatic depletion of intra-endoplasmic reticulum Ca^{++} stores. This, subsequently, leads to the activation of store-operated Ca^{++} channels (SOCCs) which are known to be responsible of massive Ca^{++} influx into the thapsigargin-activated cell. Using this model, along with the Fura-2 Ca^{++} imaging, Koss *et al.*, (2009) were able to show that genistein remarkably reduces both the amplitude and the latency of the peak response of SOCC-mediated thapsigargin-elicited Ca^{++} influx into dissociated hippocampal neurons isolated from Sprague-Dawley PND3-5 rat pups.

VI.1.4. Neurogenic Features

Under the paradigm of neurogenic signalization by isoflavones, Wang *et al.*, (2008a) have demonstrated that, even used at a 100 µM concentration in the culture medium, daidzein does not impair the viability of dissociated hippocampal neurons isolated from ED17 Wistar rat embryos. More interestingly, daidzein dose-dependently increased both axon formation and elongation. At 30 µM concentration, daidzein treatment resulted in:

1) Increased number of filopodia,
2) Increased translocation of GAP43 to the tip of elongating neurites,
3) Increased translocation of ERβ, but not ERα, to the plasma membrane, and
4) Increased phosphorylation levels of both GAP43 and protein kinase C α isoform (PKCα).

VI. 2. Discussion of Pharmacological Effects

As described above, the pharmacological investigations of isoflavone effects on hippocampal neurons, derived from a number of studies that have been using these molecules in a wide spectrum of concentrations ranging from 5 nM (for daidzein, Wang *et al.*, 2008a) to 250 µM (for genistein, Casey *et al.*, 2002). Although, all this data is interesting to consider, those that did use isoflavone in pharmacological dosing likely reachable by dietary means, will be pointed out.

Concerning the bio-electrical features of isoflavone effects, it was shown that, in both CA1 pyramidal neurons (O'Dell *et al.*, 1991) and DG granule cells (Casey *et al.*, 2002), genistein could negatively impact the expected induction of LTP that normally follows the appropriate electrical stimulation of the respective afferent pathways. However, these two studies did use very high genistein concentrations (110 µM, for the study by O'Dell and 220µM for the study by Casey). Although, in the study by Casey and colleagues, the concentration of genistein at the immediate vicinity of DG granule cells was, most probably, much inferior to 250 µM for 4 main reasons:

1) Genistein-containing solution was injected into the lateral ventricle where the cerebro-spinal fluid (CSF) may have diluted the drug,

2) A very small volume was injected (5 µl), and thereof, the CSF dilution effect of the injected drug solution should not be neglected,

3) The CSF is a circulating biological fluid, and thereof, some quantity of the drug might have been carried away from the lateral ventricle before having reached the hippocampal formation, and, then, the DG granule cells, and

4) The waste inherent to such intervention (intra-cerebro-ventricular injection, icv) in whole animal.

Quite interestingly, using genistein at a much lower concentration (20 µM) in the hippocampal slice culture medium, results in a reduction of induced LTP in the CA1 pyramidal neurons of young rats and an almost complete abolition of induced LTP in CA1 pyramidal neurons of aged rats (Shankar et al., 1998).

Nonetheless, the concentrations reported previously, most likely, would not be reached in the circulating plasma, under the conditions of nutritional consumption of even a highly dosed isoflavone-containing diet or supplementation. In addition, brain slicing, in an unrealistic fashion, exaggerates the exposure of cells to chemicals, and most probably amplifies the biological cell response. Because blocking LTP induction by blocking the associated NMDA receptor signalization impairs both spatial reference memory and verbal memory (Rowland et al., 2005), the argument in favor of our assertion is that animals exposed to highly dosed isoflavone-based diets are still able to learn and to retain spatial information in hippocampus-challenging behavioral tasks (Lund and Lephart, 2001a, b; Lund et al., 2001;Y. Lee et al., 2004; Luine et al., 2006; Monteiro, 2008). Moreover, E_2 has been shown to increase both the CA1 neuronal excitability through a non-NMDA-dependent mechanism (Wong and Moss, 1992); and the magnitude of LTP in the CA3-CA1 synapses through a genomic mechanism involving an up-regulation of the NMDA receptor subunit, NR2B (Foy et al., 1999; Smith and McMahon, 2005; 2006; Smith et al., 2009); or, at least, through the translocation of NR2B subunit to sites of active synapses (Snyder et al., 2010) and/or through post-transcriptional modulation of NMDAR1 subunit (Gazzaley et al., 1996). The use of high doses of genistein in the two studies mentioned earlier, was obviously intended to induce a pharmacological inhibition of hippocampal PTKs in order to gain insights into the mechanisms of LTP induction and, thus, should be viewed as mechanistic investigations. Indeed, the respective authors did report robust genistein-mediated inhibitions of pp60c-rsc(+), protein tyrosine kinase, activity (O'Dell *et al.*, 1991) and ERK phosphorylation (Casey *et al.*, 2002). In keeping with this, E_2-treatment of hippocampal slices also induced LTD and an ERα-mediated, increase of spine density on CA1 neurons and a decrease of thorn density on CA3 neurons, most likely, through the activation of ERK/MAPK pathway (Ogiue-Ikeda *et al.*, 2008). Taken together, these findings may suggest that the inhibition of hippocampal PTKs is unlikely in humans consuming isoflavone-containing foods, since the reported plasma concentrations do not exceed (10 μM) (Mathey *et al.*, 2006). Additionally, the pharmacological effects of high doses of genistein on the bio-electrical activity of hippocampal neurons were also seen as:

1) an inhibition of IGF-1-induced, AMPAR-mediated increased excitability (220μM, Ramsey *et al.*, 2005),
2) an alleviation of mGluR-mediated inhibition of I_{AHP}, (100 μM, Abdul-Ghani *et al.*, 1996),
3) a blockade of 4βPMA-mediated potentiation of NMDA-activated currents (50 μM, Lu *et al.*, 1999),
4) a reduction of depolarization-coupled HVA Ca++ currents (100 μM, Potier *et al.*, 1999), and
5) a potentiation of the ACh-evoked currents (100 μM for Charpantier *et al.*, 2005; 10 μM for Cho *et al.*, 2005).

Quite strikingly, completely opposite effects have been reported for much lower concentrations of E_2, in different intra-cellular signalizations taking place in brain neurons. Indeed, E_2 was shown:

1) to elicit induction of IGF-1 mRNA expression in immortalized hippocampal neurons (Shingo and Kito, 2003),
2) to activate mGlu receptors through an ER-dependent mechanisms (Boulware and Mermelstein, 2009) and to decrease the amplitude of I_{AHP} currents, most likely, through an inhibition of L-type Ca^{++} channel (Kumar and Foster, 2002),
3) to enhance an NMDA-dependent induction of hippocampal LTP (Foy *et al.*, 1999; 2008; Smith and McMahon, 2005; 2006; Smith *et al.*, 2009)
4) to bind directly and, consequently, to potentiate L-type voltage-gated

Ca^{++} channel (VGCC) currents in embryonic ED18 primary cultured hippocampal neurons (Sarkar *et al.*, 2008),

5) to inhibit nicotine-evoked $\alpha_4\beta_2$nAChR currents (Damaj, 2001) although an allosteric effect-mediated potentiation of ACh-evoked human $\alpha_4\beta_2$nAChR currents was reported for E_2 in both *Xenopus* oocytes and HEK cells (Curtis *et al.*, 2002).

Altogether, these results confirm that the findings obtained with high doses of isoflavones, and more particularly of genistein, in hippocampus-derived neuronal systems, are more relevant to PTK-inhibiting signalizations assimilated to anti-estrogenic effects rather than E_2-mimicking signalizations.

We also reported data depicting physio-molecular alterations that were triggered by isoflavone molecules. The data did, for example, document the mechanistic features of channel trafficking in hippocampus-derived cell systems. Here again, a wide range of isoflavone concentrations have been used. More specifically, genistein was capable of blocking insulin-elicited AMPA receptor internalization with relatively a lower concentration than those used to impact the bio-electrical activity (1 μM, Beattie *et al.*, 2000). This dose that is, probably, reachable under normal nutritional conditions, might affect some AMPA-related neuronal excitability and/or LTP expression in subjects exposed to dietary/supplemental isoflavones. However, such investigations still have to be conducted under nutritional isoflavone content-controlled experimental conditions.

Finally, concerning their well established characteristics as robust neuroprotective molecules in numerous cell systems, we described the wide extent of isoflavone protective effects upon hippocampal neurons that have been challenged by different cell stressors. These life-saving features of isoflavones upon hippocampal neurons might promote the integrity of hippocampal neuronal networks and, consequently, that of the upon-grafted learning and memory functions.

Furthermore, one interesting aspect of isoflavone effects on hippocampal neurons consisted in their ability to operate as neurogenic/neurotrophic signaling molecules. Indeed, daidzein was capable of triggering axonal outgrowth in dissociated hippocampal neurons at 30μM concentration (Wang *et al.*, 2008a). Although, such a dose is, likely, still out of reach to a dietary/supplemental isoflavone exposure, this effect of daidzein resembles those exerted by estradiol on a variety of brain neurons (De Lacalle 2006). Therefore, daidzein should be tested on such outcomes at lower doses for longer exposures.

Chapter VII

Pharmacological Effects of Isoflavones in Non-Hippocampal Cells

In the present section, we provide some data describing the effects of isoflavones in non-hippocampal neuronal systems. Because gaining some more insights into the effects of isoflavones in non-hippocampal, neural tissues may be helpful in highlighting further the mechanistic regulations taking place in the hippocampal formation. We also provide data depicting the effects of isoflavones that have been studied in non-neuronal cells. This might be helpful in broadening our view angle on the mechanistic features of isoflavone signalization.

VII. 1. Isoflavone Signalization in Neuronal Cells

Paillart *et al.*, (1997) have reported that genistein binds directly and inhibits voltage-sensitive sodium (Na^+) channels in rat fetal brain neurons and cerebellar granule cells ($IC_{50} = 60$ μM). Also, daidzein ($IC_{50} = 195$ μM) was able to mimic the inhibitory effect of genistein on Na^+ channel activity. Genistein (50 μM), but not daidzein (50 μM), reversibly reduces the slope conductance of the γ-amino-butyric-acid-A (GABA-A)-mediated currents in cultured dorsal medulla neurons from Wistar rat foetuses (Wan *et al.*, 1997). In a model of βamyloid-induced cell death, genistein (50 μM) was reported to have no neuroprotective effect upon cultured cortical neurons from Sprague-Dawley PND1 rat pups (Wang *et al.*, 2001). In addition, genistein (50 µM) was found to be rather pro-apoptotic on cultured cortical neurons from ED18 fetuses (Linford *et al.*, 2001). Using retinal rod photoreceptors and olfactory receptor neurons from tiger salamanders, and transfected Xenopus oocytes, Molokanova *et al.*, (1999, 2000) were able to show that genistein at doses as low as 10 µM and up to 1000 µM dramatically attenuates the apparent affinity of cyclic nucleotide-gated (CNG) channels for cyclic-guanosine-mono-phosphate (cGMP), and inhibits the maximal current through CNG channels. Interestingly in the absence of ATP, the inhibition is already observed, although not complete, from 1 µM exhibiting a nice dose-response effect. A human embryonic carcinoma-derived cells, namely hNTera-2/CD1 (also called NT-2) cells, exhibit human neuronal progenitor cell-like characteristics after continuous exposure to retinoic acid (under these conditions, these cells undergo neuronal differentiation). In this model, genistein was shown to induce neuronal differentiation and neurite outgrowth at low dose (5µM) but not at high doses (10 and 15 µM). It also induced ERK1/2 phosphorylation in an irreversible way and a dose dependent manner and up-regulated protein expression for cyclin-dependent kinase (Cdk5), protein-21 (p21) and N-cadherin indicating a positive role in neuronal differentiation rather than on neuronal proliferation (Hung *et al.*, 2005). In human SH-SY5Y neuroblastoma cells, which are known to express both ERα and ERβ, genistein (at 1 and 10 µM) displayed pro-survival effects against βamyloid-induced neuronal death and DNA fragmentation (Bang *et al.*, 2004). These effects of genistein, that were abrogated by the ER antagonist, ICI 182,780, did not seem to be mediated through a blockade of protein tyrosine kinase activity but rather through an ER-mediated genomic action (Bang *et al.*, 2004). Bartlett (1997) have demonstrated that under a low concentration of nerve growth factor (NGF) in the culture medium (100 fg/mL) and even used at a low dose (2.5µM), genistein was able to promote efficiently both neuronal survival and neurite outgrowth in cultured embryonic chick sensory neurons. However, in a model of βamyloid-induced cell death, genistein (50 µM) was reported to have no neuroprotective effect upon cultured cortical neurons from Sprague-Dawley PND1 rat pups (Wang *et al.*, 2001). On the contrary, genistein was found to be rather pro-apoptotic on cultured cortical neurons from ED18 foetuses (Linford *et al.*, 2001). Schreihofer and Redmond (2009) addressed the neuroprotective ability of isoflavones, genistein, daidzein and equol, in cultured primary cortical neurons taken from Long-Evans rat ED18 fetuses. In that case, the neuroprotective ability of theses isoflavones was investigated under ischemia-like conditions, namely, glutamate-insult, thapsigargin-induced apoptosis, hypoxia and oxygen-glucose deprivation (OGD). The main findings were as follows: (1) genistein, daidzein, and equol, at low concentrations (1µM), were all capable of protecting embryonic cortical neurons from ischemic insult-induced apoptosis through an ER-mediated mechanism; (2) genistein, most likely, operated through a MAPK/PI3K-dependent mechanism; (3) daidzein did act through a PI3K-dependent mechanism; and, (4) equol, too, may have, possibly, operated through a PI3K-dependent mechanism. Somponpun and Sladek (2002) have reported that genistein at 0.1 µM is an efficient inhibitor of NMDA-stimulated AVP and Oxyt release from explants of the hypothalamo-neurohypophyseal system, probably, through an ER-mediated mechanism.

VII. 2. Isoflavone Signalization in Non-Neuronal Cells

As a widely expressed neurotrophin, the brain-derived neurotrophic factor (BDNF), play a crucial role in the neuronal signalization throughout the central nervous system and, more specifically, within the hippocampal formation neuronal networks (Cowansage et al., 2009). BDNF activates tyrosine kinase (TrK) receptors, such as the TrKB receptors. In turn, TrKB receptors are able to trigger intra-cellular/membrane signalization by initiating a cascade of protein phosphorylations. Rogalski et al., (2000) reported that genistein (100 µM) blocks BDNF-activated, TrKB-mediated inhibition of the G protein-activated inwardly rectifying K+ channels (Kir.3) as in a transfected Xenopus oocyte model. Using ERα-positive MCF7, ER-negative/GPR30-positive SKBR3, GPR30-negative MDA-MB231 and GPR30-negative BT-20 breast cancer cells, Maggiolini and colleagues (2004) did demonstrate that genistein (1 µM) is highly efficient in eliciting c–fos, pS2 and cathepsin D gene expression through both ERα-dependent and ERα-independent mechanisms. The ERα-independent mechanism involved GPR30-mediated transduction of genistein signal through Gβγ-associated activation of ERK1/2 pathway. Vivacqua et al., (2006) have investigated genistein signalization in human WRO cell line, human anaplastic FRO and ARO thyroid tumor cell lines and in HeLa cells. These authors reported that the genistein-enhanced proliferation rate obtained at 0.1 µM was associated with an up-regulation of the expression levels of cyclin A, cyclin D1, Bcl-X$_L$ and c-fos mRNAs, while the expression levels of progesterone receptor or pS2 remained unaffected by genistein (1 µM for one hour). A c-fos promoter lacking the ERE-responsive element was still able to mediate transactivation by genistein (1 µM), most probably, through a MEK/ERK/PI3K/MAPK signaling pathway. Interestingly, genistein-elicited responses were completely abrogated when the cells were pre-treated with a GPR30 antisense oligodeoxynucleotide (AS-ODN). Under the paradigm of rapid signalization, the GPR30 is capable of translating, within few seconds-to-minutes time laps, the extra-cellular E_2 signal into an intra-cellular one in a number of cells. Indeed, Thomas and Dong (2006) used a human HEK293 cell line devoted of ERα, ERβ, and GPR30 to investigate the putative rapid signalization by genistein (0.2 µM) in GPR30-transfected HEK293-cells. They showed that genistein (0.2 µM) binds with high affinity ($1/10^{th}$ of that of E_2) to plasma membranes prepared from GPR30 expression vector-transfected cells. The binding of genistein to the plasma membranes elicited a robust production of cAMP, most likely, throughout the activation of a membrane-associated adenylyl cyclase system.

Quite interestingly, another aspect of isoflavone signalization is exemplified by the fact that these molecules are capable of modulating the endogenous estrogen metabolism. Indeed, in a cell culture model of hepatic HepG2 cells, and oppositely to E_2, genistein was shown to increase, dose-dependently (0.1 to 10 µM), both the enzymatic activity level and the gene expression level of the aromatase, P450-CYP19 (Ye et al., 2009). Albeit, it was not statistically significant at the dose of 0.1 µM, the tendency of the aromatase activity to increase was already obviously seen at this concentration (Ye et al., 2009). It became manifest at the concentrations of 1 µM and 10 µM (Ye et al., 2009). This effect that could not be triggered by E_2, involved the activation of the intra-cellular PKCα/P38/ERK1-2 signalling pathways that lead to the ultimate activation of CREB-dependent gene regulation (Ye et al., 2009). Similar effect was obtained by genistein (1 nM) in human endometrial stromal cells (Edmunds et al., 2005). In contrast, biochanin A was shown to down-regulate aromatase P450-CYP19 activity in CYP19-transfected MCF-7 cells and the mRNA expression levels in an ER-negative, breast cancer cell system, SK-BR-3 (Wang et al., 2008a). Similar inhibitory effects were obtained by genistein (10 µM) and daidzein (10 µM) or by the combination of genistein (100 nM), daidzein (100 nM) and biochanin A (100 nM) on human granulosa-luteal (GL) cells (Rice et al., 2006). In the same context, isoflavone-triggered inhibition of aromatase and 3β-hydroxysteroid dehydrogenase (HSD) activities was obtained by genistein (10 µM), daidzein (10 µM), biochanin A (10 µM) in human granulosa-luteal (GL) cells (Lacey et al., 2005). The fact that most of these concentrations, that did efficiently modulate the activity of the steroid metabolizing enzymes, are likely reachable under nutritional conditions, suggest that dietary/supplemental isoflavones may be capable of triggering some E_2-non-mimicked, own specific effects in different organs of the body and, probably, in some brain neurons, as well. Although, a previous study that did examine the dietary effects of a isoflavone-rich diet (Phyto-600) on the enzymatic activity of aromatase, could not report any significant change in the preoptic area/medio-basal hypothalamus of Phyto-600 fed male and female rat brains (Lephart *et al.*, 2001). Therefore, isoflavone-mediated regulations of neurosteroidogenic enzyme activities remain to be addressed in brain cells and, more particularly, in the hippocampal formation neurons.

Chapter VIII

Physio-Molecular Impact of Isoflavones on Brain Physiology

The studies presented here deal mainly with dietary exposure for animal models. To help interprete the respective results, a tentative correlation between animal and human exposure is given here. First, according to Adlercreutz *et al.* (1993) a Japanese man consume between 1 and 2 kg of food a day containing between 45 and 60 mg isoflavones. Second, a rodent needs between 5 and 10 times more isoflavones to reach the same plasma bioavailable isoflavone levels than man (Xu *et al.*, 1994; Izumi *et al.*, 2000; Picherit *et al.*, 2000). This means that a 10 mg/kg of body weight isoflavone exposure in rodent is roughly equivalent to 1 or 2 mg/kg of body weight in humans. Besides impacting the functioning of both hippocampal and non-hippocampal neuronal systems, and non-neuronal cell systems, as well, isoflavones have also been shown to deeply impact the brain physiology. This impact may probably affect indirectly the hippocampal formation function.

Lephart *et al.*, (2000) evaluated the impact of a diet highly enriched with soy isoflavones (Phyto-600, i.e., 600 µg/g of diet, diet i.e., in man between 60 to 120 mg/kg of food roughly twice the exposure reported in Japanese people) on brain physiology of Sprague-Dawley male rat. They showed that Phyto-600 fed rats displayed increased levels of isoflavones in circulating plasma. The bioavailable doses were 2465 ng/mL which is about 10 µM and can be reached in certain humans with 50 mg isoflavones twice a day in the morning and the evening (Mathey *et al.*, 2006). Lephart *et al.*, (2000) did further show that Phyto-600 diet results in: (1) increased levels of isoflavones in medio-basal hypothalamus/preoptic area (69,9 ng/g); (2) unaffected levels of enzyme activity of aromatase and 5-α-reductase, two androgen-metabolizing enzymes, in medio-basal hypothalamus/preoptic area; (3) decreased mRNA expression levels of the neuroprotective Ca^{++}-binding proteins, calbindin and calretinin in the amygdala; and, (4) decreased mRNA expression levels of calbindin, but not calretinin, in mediobasal hypothalamus/preoptic area. In the adult male rat frontal cortex, isoflavone-enriched diet has been associated with both decreased expression levels of calbindin mRNA (a pro-survival factor) and increased expression levels of cyclooxygenase-2 (an inducible inflammatory factor). The same research group (Bu and Lephart, 2007) did show that Phyto-600 fed male rats displayed 22% decrease in the number of ERβ-immunopositive neurons in the anteroventral periventricular nucleus (AVPV), most likely due to an increased rate of apoptosis occurrence as evidenced by the increased co-expression of TUNNEL immuno-staining and ERβ-IR within AVPV neuronal population. Pan *et al.* (1999b) studied the effects of a soy isoflavone-enriched diet (equivalent for a woman, to a daily intake of 150 mg of isoflavones) in OVX Sprague-Dawley retired breeder rat brains. They showed that, after 8 weeks on diet, dietary isoflavones were able to up-regulate the expression levels of BDNF mRNA in the frontal cortex of OVX rats. This effect resembled that of a control diet with added E_2 (equivalent to 2mg/day) (Pan *et al.,* 1999a). On the contrary, in adult male hooded Lister rats, a soy isoflavone-containing diet (150 μg isoflavones/g of diet) was associated with a down-regulation of BDNF mRNAs in the CA3 and CA4 fields of the hippocampal formation and also, in the cerebral cortex, although the expression levels of GAD67 mRNAs and GFAP mRNAs were not affected by the diet in the studied regions (File *et al.,* 2003). More specifically, a genistein-enriched (1250 ppm) diet was associated with increased levels of hypothalamic contents in arginin-vasopressin (AVP), but not β-endorphin (Scallet *et al.,* 2003). In Long-Evans male rats exposed to Phyto-600 diet, as compared to those fed an isoflavone-free diet, the expression levels of BAD protein (a proapoptotic factor) were: (1) increased in the amygdala; (2) decreased in the mediobasal hypothalamus and frontal cortex; and, (3) unaffected by the diet in the cerebellum and the hippocampus. Whereas the expression levels of βIII-tubilin (an early marker of neuronal differentiation/survival) were: (1) increased in the amygdala, hippocampus, mediobasal hypothalamus, and frontal cortex; but, (2) unaffected in the cerebellum. Under the paradigm of aversive conditioning, day-old chicks were tested in a discriminated passive avoidance task. It was found that intra-striatal injection of genistein produces a strong amnesic effect on passive avoidance learning (Whitechurch *et al.,* 1997) probably, through an anti-protein tyrosine kinase mechanism. Y. Lee and colleagues (2004) have compared the effects of 2 isoflavone-dosed diets (0.3g of isoflavones/kg of diet and 1.2g of isoflavones/kg of diet; one plausible doses for human exposure and the other about twice the maximum recorded in Asia) to a third isoflavone-free control diet, on age-related impairment of hippocampal functioning and brain cholinergic signalization in aged Sprague-Dawley male rats. These authors have reported that isoflavone-enriched diets decreased acetylcholine esterase (AChE) activity and increased ChAT activity in hippocampal/basal forebrain/cerebral cortical areas, and increased ChAT-IR in both medial septum and hippocampus. All these effects were not seen altogether in the same diet group. In OVX rats, both E_2-treated (3.9 mg/Kg b.w. very high dose) and high soy isoflavone-rich diet (526.9 mg/Kg largely over the human consumption) fed group did display both lower and higher levels in, respectively, hippocampal ACh esterase (AChE) activity and basal forebrain ChAT-IR as compared to the two other groups that have been fed with a low soy isoflavone-rich diet (263.4 mg/Kg about 2 or 3 mg/kg b.w. for a human being) or a soy isoflavone-free control diet (Lee *et al.*, 2009). Soy isoflavone-triggered lowering of both hippocampal and frontal cortex AChE activities was also seen in hypercholesterolemic mice (Liu *et al.*, 2007). These findings show that at high doses, soy isoflavones are capable of impacting positively, the cellular metabolism of hippocampus-projecting medial septo-basal forebrain cholinergic neuronal system. Therefore, dietary soy isoflavones might have some inherent potential for the alleviation of age-related degeneration of ACh-synthesizing neurons, and maybe also in human subjects. However, because only E_2 did increase the ChAT enzyme activity in the hippocampus of OVX rats (Lee *et al.*, 2009), it seems evident that isoflavones, or at least the soy isoflavone extract used herein, would not trigger all the beneficial effects of native E_2 in the exact same brain areas, not to mention when considering different brain structures. This is a red flag to be waved in the face of those who want to replace estrogen hormone therapy with isoflavones, not to mention the pro-apoptotic/pro-inflammatory effects triggered by these substances in some specific brain regions (discussed earlier). Nevertheless, this assertion must be tempered by the fact that we do not know much about the bioavailability/signalizing effects of isoflavones in the human brain and, more particularly, at the level of the hippocampal-projecting cholinergic system.

Chapter IX

Effects of Dietary/Supplemental Isoflavones on Cognitive Processes

In the present section, we report on data that used isoflavones as dietary and/or supplemental products for their putative ability to impact cognitive brain function. By doing so, we hope to get the largest view possible on the signalization potential of these molecules under normal conditions of daily nutritional use. Firstly, we are going to summarize these data according to the model examined. Secondly, we are going to discuss the main resulting findings in the light of what is known about the physiology of the hippocampal formation.

IX. 1. Description

In the present section, we report on data that used isoflavones as dietary and/or supplemental products to deepen our knowledge of their impact on cognitive brain function. Indeed, dietary/supplemental isoflavones have been investigated in: (1) different strains of rat (Long-Evans, Sprague-Dawley and Wistar rats); (2) non-human primates; and, (3) young, middle- aged, and older human subjects. Both genders have been documented in humans in nutritional isoflavone studies. In these studies, the assessment of the cognitive function used well established psychological tasks and mood rating questionnaires. In non-human subjects, however, specific apparatus-based behavioral tasks were called for to gain some objective insights into the learning and memory processes (see Table 2 for a complete summary).

IX. 1.1. Rodents

It is no surprise that, concerning the investigation of isoflavone signalization in whole animals, most studied have been conducted in rodent species (Tables 2.1-3). For example, the impact of a lifelong exposure to a isoflavone-rich diet (Phyto-600, containing 600 µg of isoflavones/g of diet, roughly equivalent in men to 60-to-120mg of isoflavones/day) on memory function was evaluated in Long-Evans male and female rats in the 8 arm-radial-maze. It was reported that, in reference memory-related tasks, the isoflavone-rich diet improves the performance of both female rats and *in utero* pharmacologically-feminized male rats, whereas it impairs the performance of intact male rats (Lund and Lephart, 2001a; Lund *et al.*, 2001). By contrast, the same diet produced anxiolytic behavior in both genders when rats were tested in elevated-plus maze (Lund and Lephart 2001b). Using OVX Sprague-Dawley female rats (2 months of age), Luine and colleagues (2006) did compare the relative effects of a high soy isoflavone-containing diet (810 µg/g of diet, roughly equivalent in men to 80-to-160mg of isoflavones/day) and a low isoflavone-containing diet on the memory function by mean of object placement and object recognition tasks. These authors were able to demonstrate that the animals fed the high-isoflavone-containing diet spent more time exploring the same object at the new location (improved spatial memory), but not the new object at the same location (unaffected object recognition memory). However, as measured in an elevated-plus maze, the instinctively driven exploratory behavior did not differ between the 2 diet groups. Using OVX adult female Wistar rats (87-117 days of age), Monteiro *et al.*, (2008) compared the relative effects of a soy isoflavone-rich diet (containing 250 µg genistein/g of soy proteins and 230 µg daidzein/g of soy proteins i.e., roughly equivalent in men to 50-to-100 mg of isoflavones/day) and a soy isoflavone-free diet on the spatial memory function as assessed in the Morris water maze. The results showed that when animals were put on the isoflavone-rich diet for 30 consecutive days after OVX, all the impaired cognitive parameters seen in the isoflavone-free diet-fed OVX-group were re-adjusted to the levels of those seen in ovary sham-operated group. Indeed, the

OVX resulted in: (1) an increased latency to find the platform; (2) an increased latency to cross the platform; (3) an increased time spent in the opposite quadrant to target; and, (4) a decreased time spent in the target quadrant. The isoflavone-enriched diet was able to improve all these 4 parameters and, thus, to rescue the impaired spatial memory function seen in OVX rats. On the contrary, a 60 days pre-treatment (diet initiated 60 days before the rats were OVX) with dietary soy isoflavones could reverse only one cognitive parameter, namely the increased latency to the hidden platform. Y.Lee *et al.* (2004) have compared the effects of 3 diets: (1) a low isoflavone-dosed diet (0.3g of soy isoflavones/kg of diet, i.e., close to Japanese diet in humans); (2) a high isoflavone-dosed diet (1.2g of soy isoflavones/kg of diet, i.e., twice or fourth times in the classical Japanese diet); and, (3) an isoflavone-free control diet, on age-related impairment of hippocampal-dependent spatial memory function in Sprague-Dawley male rats. These authors addressed the aging hippocampal function in a water Morris delayed matching-to-place (DMP) task. They reported a significant improvement in the performance of middle aged male rats (14 months of age) put on the low isoflavone-dosed diet (0.3g/kg) as compared to rats put on the 2 other diets. On five stages of acquisition trials, the animals on the low isoflavone-dosed diet swam faster (on stages 2, 3, and 4) than the animals on the other diets. Interestingly, there was no significant difference in the performance between the 2 groups put on high isoflavone-dosed diet (1.2g/kg) or isoflavone-free diet. However, this same research group (Lee *et al.*, 2009) have reported that, in OVX rats, the performance in the Morris water maze was rather improved, as well as with the E_2-treated OVX rats (3.9 mg/Kg b.w.), in the OVX rats that had been on a high soy isoflavone-rich diet (526.9 mg/Kg; four or eight times the Japanese diet) as compared to those kept on a low soy isoflavone-rich diet (263.4 mg/Kg; twice or four times the Japanese diet) or a soy isoflavone-free control diet.

IX. 1.2. Non-Human Primates

To our knowledge, only one study did address the cognitive effects of soy in a non-human primate model (Table 3). In this study, Golub and colleagues (2005) investigated the existence of putative effects of a soy-rich infant formula given for four consecutive month duration on general health parameters, including cognitive development, in Rhesus monkey infants. The isoflavone concentration in the soy-based infant formula was not documented.

A second experimental group was given a soy-rich infant formula with added manganese (Mn). The third experimental group (control subjects) was given a cow's milk based formula. To assess the cognitive brain function in the developing monkeys, four behavioral tests were regularly administered to subjects, namely: (1) object discrimination learning and reversal test; (2) delayed non-match to sample test; (3) position learning and reversal test; and, (4) reward delay test. Up to 18 months of age, none of the 4 measures resulting from these 4 tests was significantly affected by the soy-rich infant formula in the developing monkeys. Moreover, no alteration of the cerebro-spinal fluid (CSF) contents in catecholamine (dopamine, DA, or serotonin, 5HT) metabolites could be found in soy fed monkeys.

IX. 1.3. Human Subjects

In the human studies of isoflavone impact on cognitive abilities, neuropsychological tests have been proposed to participants, most of the time, at baseline and at the completion of the dietary/supplemental intervention period. These tests were designed to evaluate the performance of participants in attention and memory tasks. They were also designed to evaluate both the executive functions and the mood states of volunteers. To help interpreting these data, a sample of behavioral evaluations of isoflavone impact on cognitive abilities is summarized in Tables 4.1-5. When available, the information concerning the brain territories activated under those cognitive tests and the source of the tests is also given. In behavioral studies on humans, the duration of exposure to isoflavones ranged from 12 weeks to 12 months. All the studies that did test an active treatment, were designed as double-blind, placebo-controlled, randomized studies (Tables 5.1-8).

Epidemiological Studies

A cross-sectional correlation study between dietary intakes of lignans or isoflavones, using a food-frequency questionnaire, and brain functioning was carried-out in post-menopausal Dutch woman (age = 60-75 years). Episodic memory was evaluated by the Rey's Auditory Verbal Learning Tests (Rey immediate recall, Rey delayed recall, and Rey recognition tests). Visual memory was assessed by the Doors Test. The appreciation of short-term/working memory materialized as a measure of the performance in the digit span forward/reverse tests. Verbal fluency was addressed by the letter fluency and category generation tests. Attention abilities and mental flexibility were assessed by the trail making (A1/A2/B) tests. The Boston Naming Test was used to evaluate the verbal competence and the semantic retrieval memory. The evaluation of cognitive/perceptual speed materialized as a measure of the digit symbol substitution test. Finally, the Dutch Adult Reading Test was used in order to assess the verbal intelligence quotient. The outcomes in this study did show significant correlation between the daily intake of lignans and improved performance in both attention/mental flexibility tasks (trail-making A1 and A2) and cognitive/perceptual speed task (digit symbol substitution task). However, the neuropsychological assessment results could not be correlated to any feature of the dietary isoflavone intake of participants (Kreijkamp-Kaspers *et al.*, 2007). However, in this population the isoflavone intake was shown to be rather low (0.18, 0.34, 2.99, and 14.64 mg/day according to the different quartiles considered). Hogervost and colleagues (2008) have conducted a cross-sectional correlation study where they did compare the daily consumption of 2 soy-derived products: tofu (a "soy milk" preparation, precipitated with $MgCl_2$) and tempeh (a fermented soy preparation) to the memory performance. The results showed that high tofu intake was associated with low scores in the memory tests, whereas high tempeh intake was independently related to higher scoring in memory tests. Another correlation study carried out by Huang *et al.*, (2006) in middle-aged (age = 42-52 years) Asian women looked for any parallel between the estimated daily intake of isoflavones (as assessed by a food frequency questionnaire) and cognitive performance. In this study, the estimated median intake of genistein ranged from 3.5 µg/day-to-6.8 µg/day. No significant link between the isoflavone consumption profiles and the cognitive performances could be found within the experimental groups. Two epidemiological studies now report a higher risk of dementia in men with high daily tofu consumption, more than twice a week (White *et al.* 2000; Hogervorst *et al.*, 2008). Tofu was found by Coward *et al.*, (1998) to contain 32 mg of isoflavones per 100 g of weigh (one serving). These studies are in accordance with that of Lund and Lephart (2001a) and Lund *et al.*, (2001) mentioned previously which showed altered memory performances in male rats under phyto-600 diet.

Interventional Studies

Casini *et al.*, (2006) have conducted a double-blind, placebo-controlled, cross-over study (12 months duration: 6 months on the active treatment or placebo followed by 1 month wash-out and then by 6 months on the alternative treatment) where they provided soy isoflavone-containing tablets (60 mg/day with: 40-45% genistein, 40-45% daidzein and 10-20% glycitein) to post-menopausal woman (age = 44-54 years) and assessed the cognitive function of participants at baseline and after each treatment. The results showed that among the 23 tests administered, 6 tests were found to be positively affected by the isoflavone supplementation. The recalling of pairs was the only test improved that concerned the memory function. The 5 other tests significantly improved by the supplemental isoflavones, concerned the mood states of the participants. File and colleagues (2001) examined the effect of a 10-week-treatment with a soy diet containing 100 mg of soy isoflavones on cognitive brain function in young men and women aged 20-30 years. The authors assessed the brain function of participants at baseline and after treatment, by using psychological tests of: (1) attention (digit symbol substitution test, digit cancellation test and paced auditory serial addition test); (2) episodic memory (immediate story recalling test, delayed recalling test and delayed recalling of pictures); (3) semantic memory (category generation tests); (4) frontal lobe function (verbal fluency test, rule shifting and reversal test and planning ability test); and, (5) mood rating (visual analog scale-from the Bond and Lader scales). In this study, dietary isoflavones did not affect the attention abilities of participants. They did improve, however, the performance of participants in those tests that were related to short-term memory (immediate recall test and delayed matching to sample test) and to long-term memory (delayed recalling (> 20 min) of common object pictures recalling). In addition, both mental flexibility (in both male and female subjects) and planning ability (in female subjects, only) were also among the cognitive functions significantly improved by isoflavones. By contrast, the performance scored in the letter fluency test, was improved in women, whereas it was impaired in men. Finally, the subjects under the active treatment did rate themselves as more restrained and more relaxed than the subjects that were under the control diet. A double-blind, placebo-controlled study conducted by Duffy and colleagues (2003) did investigate, in post-menopausal women (age = 50-60 years), the effects of soy isoflavones supplements (60 mg/day, for 12 weeks) on cognition. Episodic memory, semantic memory, frontal lobe function, sustained attention and mood were all rated using appropriate psychological tests. The outcomes of this study revealed that soy isoflavone supplements improved the performance in the following tests: (1) the test of delayed recall of pictures; (2) the paced auditory serial addition test; and, (3) the test of rule shifting and reversal. All the other parameters arising from the battery of tests used in the study remained unchanged between the group under the active treatment and the group under placebo. Kreijkamp-Kaspers and colleagues (2004) did address the impact of 12 month intake of an isoflavone-rich soy protein-based diet containing (genistein = 52 mg, daidzein = 41 mg, glycitein = 6 mg) on the cognitive abilities of a post-menopausal women group (aged, 60-75). The assessment of the cognitive function that was conducted as described above in Kreijkamp-Kaspers *et al.,* (2007), did not show any significant effect on cognitive performance. Fournier *et al.,* (2007) did report that introducing soy milk (with approximately, 31 mg of daidzein, 37 mg of genistein and 4 mg of glycitein) into the daily diet of post-menopausal women (between 48 and 65 years old), resulted in impaired verbal working memory (digit ordering task), whereas adding the same amount of isoflavones (30 mg of daidzein, 33 mg of genistein and 7 mg of glycitein) to cow's milk daily intake do not impair nor improve the memory function. Moreover, Ho *et al.,* (2007) who assessed the impact of daily intake of supplemental soy isoflavones (containing 80 mg of isoflavones) on brain function in post-menopausal Chinese women (age = 55-76 years) did not found any significant effects. A double blind, placebo-controlled supplemental (100 mg of isoflavones with ~ 85% daidzein and genistein) study, conducted by Gleason *et al.,* (2009) in elderly healthy men (n=15) and women (n=15) (age = 62-89 years) randomized to two groups (placebo, n=15, active treatment, n=15) did report that, within the battery of tests administered to participant at baseline and after 6 months on supplementation, 8 tests were significantly improved by isoflavones. These tests concerned visuo-spatial memory, mental construction ability, verbal fluency, and speeded dexterity. Also, this study reported that the subjects under placebo out-performed those under active treatment in 2 tests concerning the executive function. In addition, after a 12 month-long supplementation of post-menopausal women (mean age = 53 years and the last periods having occurred between 35 to 50 months earlier) with a daily intake of isoflavone-containing red clover extracts, Maki *et al.,* (2009) could not attribute any noticeable alteration of the memory function to the supplemental isoflavones. In a double blind, placebo-controlled, cross-over study, Thorp and colleagues (2009) examined the effects of isoflavone supplements (12 weeks duration with daily intake: genistein = 12 mg, daidzein = 68 mg, glycitein = 36 mg) on the cognitive functions of middle-aged (age = 49 ± 10 years) healthy men, non-habitual consumers of soya-derived products. The results showed that supplemental isoflavones were significantly associated with an overall improvement of the performance in a spatial memory-related task. Indeed, in the novel spatial working memory task used by the authors, there were 18% fewer pairs viewed, 23% fewer memory errors and 17% faster processing within the group of subjects under the active treatment. Statistical analysis revealed that there was no other cognitive task improved or impaired by the daily intake of supplemental isoflavones. A 6-month, double-blind, randomized, placebo-controlled study, that was documented by Kritz-Silverstein et al., (2003) in healthy post-menopausal women, investigated the effects of soy isoflavones (110 mg /day) on cognitive function (SOPHIA study). The authors reported a significant improvement in category fluency, verbal memory, trails B tasks in the group of women who had been on the active treatment.

IX. 2. Discussion

The compiled data reported here does establish that dietary Isoflavones can exert some significant modulatory actions on cognitive function in both humans and rodent models, but not in a non-human primate model (Rhesus monkey) in the condition tested. Some of these effects are measurably reachable processes through adequately administered behavioral tasks. Sometimes, they do correlate with specific physiological aspects of brain functioning. It is noteworthy to observe that, when it comes to their outcomes, the studies depicted here are highly conflicting with each other, especially human studies, and that the results were depending on the task examined, the age of the studied group, the gender of the considered subjects, the time of administration from last period, the duration of isoflavone treatment, the isoflavone compound administered and the form of the dietary intake (supplements vs. soy based food).

In rodent studies, dietary isoflavones did show some capacity to impact the cognitive function in intact males (Lund and Lephart, 2001a; Lund et al., 2001;Y. Lee et al., 2004), intact females (Lund and Lephart, 2001a; Lund et al., 2001) and OVX females (Lee et al., 2009; Luine et al., 2006; Monteiro, 2008). Dietary isoflavones did also impact cognitive behaviors in Sprague-Dawley (Luine et al., 2006), Long-Evans (Lund and Lephart, 2001b; Lund et al., 2001) and Wistar (Monteiro, 2008) rats. In radial arm maze tasks addressing the visual spatial memory, a life-long isoflavone-enriched diet did improve performance of both intact female rats and in utero-feminized, flutamide-treated male rats, whereas it did impair performance in control, oil-treated male rats. This implies that dietary isoflavones are:

1) Capable of reaching and impacting the brain areas that are devoted to the accomplishment of higher cognitive functions such as the hippocampus, prefrontal cortex and associative cerebral cortices,

2) Efficiently, capable of activating the neuronal network(s) sub-serving the visual spatial memory function in the upper-mentioned brain areas,

3) Sex-dependently, able to alter the behavioral expression of learning and memory processes, and

4) Capable of counter-acting a developmental sex hormone-related cognitive defect. Additionally, a much shorter time exposure (15-16 weeks) to a low-dose of dietary isoflavones was also able to improve the aging-related spatial memory performance in the Morris water maze in middle-aged (14 months old) male Sprague-Dawley rats (Y.Lee et al., 2004). Hence, the administration of dietary isoflavones can provide a beneficial relieving treatment against age-associated impairment of reference memory. However, a group of rats who has been on a high-dose of dietary isoflavones did not differ from a control group who has been on a free-isoflavone diet (Y. Lee et al., 2004) proving that the highly-dosed isoflavone diets or the highly-dosed isoflavone supplementations may not display necessarily beneficial effects on memory function. This is, most likely, due to the fact that a complex cognitive task must activate different cerebral territories that are not at the same levels of estrogen signalization. For example, such territories may differ by their:

1) Isoflavone-mediated modulation of the expression levels of ER genes,

2) Number of ER sites available for the molecular interaction with circulating isoflavone molecules,

3) Relative abundance of ER subtypes that may putatively lead to some differential ER subtype-specific signalizations,

4) Specific ultrastructural distribution of ERs that could lead to a hypothetical cell- compartmentalized isoflavone signalizations, and/or

5) Task-linked regulation of blood flow within the given brain area(s).

Elsewhere, under the paradigm of hormone replacement therapy using isoflavones, the memory performance in OVX rats was improved in the spatial reference memory by an isoflavone-rich diet in the sense that it was associated with significant decreases in both path length and escape latency monitored on the 2nd day of the acquisition trials (Lee et al., 2009). Furthermore, an isoflavone-enriched diet did increase the amount of time spent by OVX rats at exploring an old object displaced to a new position within the exploration maze (Luine et al., 2006). This finding may reflect an isoflavone-mediated improvement of the object placement memory in surgically menopausal young animals. In the same study by Luine *et al.,* (2006), the object placement memory performance of isoflavone-fed OVX rats was ameliorated in the 7th week of treatment but not in the 5th week of treatment advocating for the likelihood of duration-specific effects of isoflavones on cognitive processes. Moreover, isoflavones did not impact the memory performance in a new object recognition task administered to OVX rats at 6, 8 and 9 weeks, after the diet introduction (Luine *et al.,* 2006) implying that the modulation of the cognitive function by isoflavones depends on the type of memory needed for the resolution of the task. In the same context, an isoflavone-rich soy-based diet provided to OVX rats soon after the surgical intervention was efficiently able to rescue a hippocampus-dependent cognition, namely reference memory function that was monitored in a Morris water maze (Monteiro *et al.,* 2008). Indeed, OVX rats exposed to isoflavones did acquire the maze faster than those who were not. Also, on the day of probing, 3 parameters that can be gained from the analysis of the platform-searching behavior of rats (occupation time of the target quadrant, occupation time of the opposite quadrant and latency to cross the platform) were all improved (up-graded to the levels of parameters recorded in sham-operated rats) by the immediate exposure to isoflavones after OVX. Interestingly, in this same study, sham-operated rats that have been exposed to isoflavones for the same duration as those who were OVX, did not show any impairment in both the acquisition phase or the probing session, evidencing that nutritional adjunction of isoflavones to normal circulating levels of estrogens does not impair the hippocampal formation functioning in the ovaries retaining animals. In addition, it has been shown that soy isoflavones dose-dependently, improve spatial working memory in OVX retired breeder rats with or without added E_2, implying that dietary isoflavones do not antagonize the beneficial effects of circulating estrogens (Pan *et al.* 2000). Taken altogether, the findings described here highlights the beneficial effects of an isoflavone-rich diet in females on hippocampal-activating mnemonic tasks that have been impaired by experimental estrogen withdrawal. None of the dietary intervention resulted into an anti-estrogenic effect of isoflavones. Therefore, dietary and/or supplemental isoflavones may be of interest in the ongoing efforts for the search of a hormonal substitute for the surgically-induced or age-associated estrogen withdrawal and the resulting impairment of hippocampus-dependent cognition experienced by aged/menopausal women. In males the discrepancy observed between some results may reflect the plausible existence of a tricky dose-dependent response effect.

In non-human primates, and to our knowledge, the only study that did address the question of soy impact on the cognitive function is the one published by Golub *et al.* (2005) although the isoflavone exposure was not documented in this trial. This research group did investigate the effects of a soy formula-based diet on brain functioning in the developing Rhesus monkeys until their 18th month of age. They reported no significant impact of soy-based infant formula on cognitive abilities. However, to ascertain this possibility, the isoflavone content should have been measured and soy-fed monkey brains should have been examined at the neuroanatomical level in order to rule out any adverse occurrence such as, for example, a neurodegenerative defect.

In human studies, dietary isoflavones did improve the performance in tasks related to short-term memory (immediate recall and delayed matching to sample), long-term memory (pictures recall) and mental flexibility in young (aged, 20-30 years) subjects of both genders (File *et al.*, 2001). These results are in agreement with those reported by Gleason *et al.*, (2009) in elderly men and women (aged, 62-89 years) in whom visuo-spatial memory, mental construction ability, verbal fluency and speeded dexterity were improved by a 6-months intake of a soy isoflavone supplement. However, dietary isoflavones displayed some sex-specific effects on planning abilities that were improved in female subjects but unaffected in male subjects (File *et al.*, 2001). By contrast, some adverse effects of isoflavones can also be seen with respect to the gender of the group examined as it was the case for a letter fluency task that was improved in female subjects but impaired in male subjects (File *et al.*, 2001). In a slightly different context, supplemental isoflavones did cause a remarkable improvement of spatial working memory in healthy men aged 49±10 years (Thorp *et al.*, 2009). Similarly to the young human subjects under dietary isoflavones, short-term memory, long-term memory and cognitive processing speed were improved by an isoflavone supplementation in post-menopausal (aged, 44-65 years) women (Casini *et al.*, 2006; Duffy *et al.*, 2003). A 12 month-long daily supplementation with an isoflavone-rich red clover extract did show a null effect on cognitive function in elderly persons. However, in this specific case the bioavailable isoflavone plasma levels were very low indicating, probably, a modification in the absorption, the metabolism and/or the excretion rates of isoflavones with aging (Maki *et al.*, 2009). Similarly, a 6 month-long isoflavone supplementation did not affect cognitive function in post-menopausal women (Ho *et al.*, 2007). However, in this study the menopausal parameters were highly variable: age, time since menopause, HRT treatment or not and all these parameters would have influenced the cognitive response leading to non-statistical differences. Considering the nature of the isoflavone compound administered to participants, Fournier et al. (2007) did report an impaired verbal working memory in post-menopausal women (aged, 48-65 years) who have been on a diet containing soy milk (72 mg isflavones/day) + a placebo supplement as compared to those who have been on a diet containing a cow's milk + an isoflavone supplement (70 mg isoflavones/day) or a diet containing a cow's milk + placebo supplement. These findings suggest that the isoflavone-containing matrix (herein, soy milk) or the preparation procedure of the isoflavone-derived product, may have some detrimental actions on the brain function, regardless of the active compound tested. In addition, in cross-sectional correlation studies, high tofu intake was associated with low performance in the memory testing (Hogervost et al., 2008) and to deleterious brain functioning in old men (White et al., 2000) when tempeh was associated to memory improvement (Hogervost et al., 2008). In addition, high isoflavone-rich diets could not be associated with any improvement of, or with any adverse effect on memory testing (Kreijkamp-Kaspers et al., 2007; Huang et al., 2006). Similarly, an isoflavone-rich soy protein based diet did not impact cognitive functioning in post-menopausal women aged 60-75 years (Kreijkamp-kaspers et al., 2004). Nonetheless, the effects observed in the tofu consumer group appear to corroborate the one obtained in soy milk group. With respect to hippocampal-associated memory, discrepant results could be noticed depending on the study considered. Indeed, the performance in such memory subtype, for example, among supplemental-designed investigations, was improved in Duffy et al., (2003), Casini et al., (2006) and Gleason et al., (2009) studies but was unaffected in Ho et al., (2007) study. In keeping with this, and considering soy-based isoflavone-containing diets, hippocampus-activating mnemonic tasks were improved in File et al. study (2001) but were unaffected in Kreijkamp-Kaspers et al. (2004) study. Still, the large differences in: (1) the age of participants (20-30 years in File's study and 60-75 years in Kreijkamp-Kaspers's study, with no cycles for several years); (2) the treatment durations (10 weeks in File's study and 12 months in Kreijkamp-Kaspers's study); and, (3) the number of participants (n=27, in File's study and n=175, in Kreijkamp-Kaspers's study) may offer plausible explanations to the apparent discrepancy in the outcomes. Besides, the participants in File's study were represented by men (n=12) and by women (n=15) and the small subject number did not allow a separate statistical investigation, whereas in Kreijkamp-Kaspers's study, the participants were all postmenopausal women. Altogether, these observations suggest that dietary isoflavones might exert some: (1) exposure time-specific; (2) exposure nature-specific; (3) age-specific; and/or (4) sex-specific effects on the hippocampal-dependent memory function in humans.

Chapter X

Concluding Remarks

Cellular signalization by estrogens is a multifaceted and a widespread physiological mechanism among each and every single organ within the body including the brain. Therefore, this signalization arose as both a highly challenging research topic and a highly critical issue for human health. Consequently, due to their structural resemblance to estrogen molecules and to their natural occurrence in food including high quality proteins (for example, in soy), isoflavones have erupted into the spotlighted arena of human curiosity, having succeeded in capturing a large part of scientific, media, industrial and public attention throughout the world. Thus, in the present review, we provided *in vitro*, animal and clinical studies that did describe a plethora of effects arising from the use of isoflavones as signaling molecules. Each one of these studies could be classified in one of the following two main ensembles: pharmacological or dietary/supplemental studies. Next, we are going to recall the main result findings according to the relative nature of the study: pharmacological studies *versus* nutritional studies. In the reported effects, we will be focusing on brain neurons, and more specifically, on hippocampal neurons. We will also try to draw the main conclusions and plausible consequences as of the isoflavone impact on human cognition. Of course, keeping in mind the upper levels of plasma concentrations that can be reasonably reached by dietary/supplemental isoflavones. In parallel to our closing conclusions, we will be evoking some debatable findings that fall into a grey, shadowy area of highly controversial reporting. Also, to show that some research topics are clearly still suffering from complete lack of investigative research.

X.1. Pharmacological Studies

Herein, isoflavone molecules have been revealed as potent pharmacological substances modulating intra-cellular transduction pathways in various cell systems, and more specifically, in hippocampal formation neurons. The pharmacological studies carried out using high doses of isoflavones (> 10 μM) that are unlikely occurring concentrations in circulating plasma under the normal nutritional status, have produced signalizing effects reflecting the powerful inhibitory action of isoflavones (genistein mainly) upon the phosphorylation activity of cellular/neuronal PTKs. This inhibitory action was most likely due, at least in part, to the competitor property of this isoflavone (genistein) at the ATP-binding site of PTKs. Nonetheless, high concentrations of isoflavones produced manifest alterations of the bio-electrical neuronal activity (LTP, ion channel-sustained currents and neurotransmitter-mediated signalizations).

Albeit in a significant amount of data where the concentrations of isoflavones used would not be relevant for nutritional strategies, isoflavone molecules, nonetheless, remain robust signaling substances that must be handled with extreme caution. More specifically, when it comes to the matter of supplemental isoflavones, overdosing must be adversely forbidden and elderly people must be extremely sensitized against such practice. Also, such recommendations must be clearly advertised in supplemental product leaflets accompanying the over-the-counter products available in drug stores for public sale.

X.2. Nutritional Studies

As largely developed in the present chapter, dietary/supplemental isoflavones are powerful substances that display highly potent signalizing characteristics in various cell systems including brain neuronal systems. In particular, such isoflavones showed substantial leverage with the neuronal physiology of the hippocampal formation. In humans, the effects reported here would indicate that isoflavones at doses ranging from 60 to 100 mg a day may improve cognitive features in women when administered soon after the menopause occurrence. Although not consistently recurrent, similar outcomes have been identified for the hormonal replacement therapy using variously designed estrogens or different combinatorial configurations with co-administered therapeutical hormone candidates (for example, progesterone) along with estrogens. At lower doses (< 60mg/day), however, no effect on human cognitive abilities could be, directly or indirectly, inferred to nutritional isoflavones. In men, too, isoflavones did exhibit positive effects when administered in supplements or in soy-deriving foods, such as tempeh. Nevertheless, daily intake of high quantities of a soy-arising meal, herein tofu, seems to produce deleterious effects as those seen in two epidemiological studies. It must be mentioned here that, in the Western recipes, the 7 deepings, which are traditionally applied in the Asian recipe, are most often omitted. These Deepings associated to heating, most probably, reduce the content of the soya matrix in antinutritional compounds (urease, trypsin inhibitors, phytase, saponins, hemaglutinins, stigmasterol) and in isoflavone of traditional tofu, since isoflavones in their glycosylated forms, are soluble in water. It is unclear from the animal/human studies if high intake of nutritional isoflavones would produce positive or rather deleterious effects, since in all the studies that have shown a positive impact, either administered doses or reported plasma levels were low or medium. Furthermore, the effects of isoflavones on both the cognitive behaviors (in animal and human studies) and the brain physiology indicate that not only the hippocampus would be influenced by dietary isoflavones. Most probably, brain areas like the pre-frontal/frontal cortices, amygdala, striatum, preoptic/medio-basal hypothalamic structures, midbrain and brainstem, that are well known to be both highly enriched in ERs and functionally interconnected, are plausible target sites for some isoflavone-mediated signalizations (Figure 2). These signalizations, therefore, would transpire behaviorally as hippocampus-involving improvement/impairment of the memory function.

Isoflavones raised the hope of being used as alternative brain-protecting drugs to hormone replacement therapy and in aging/menopausal-associated dysfunctions, but an eye still have to be pointed out on their proliferative effects in estrogen-dependent cancers. However, as reported enclosed, isoflavones do trigger some alarming effects on neuronal survival, neuronal excitability, brain physiology, cognitive function and, thus, these molecules still have to be considered with wise caution. Hence, little skepticism must and is well still around as has been reportedly published in the scientific literature and more specifically in expert opinion-reporting reviews.

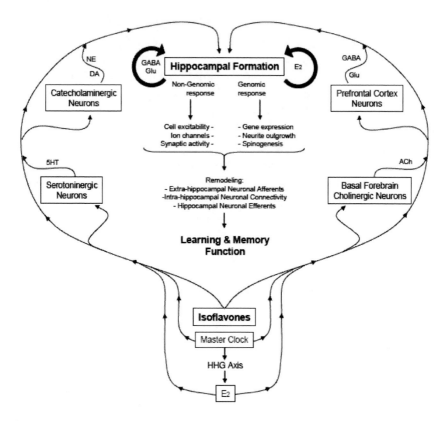

Figure 2. Putative signaling mechanisms by dietary/supplemental isoflavones in mnemonic brain structures. Dietary/supplemental isoflavones may impact hippocampal formation neurons directly and indirectly, through monoaminergic systems and prefrontal cortex neurons. Nutritional isoflavones may act through genomic and non-genomic mechanisms to re-modulate intra/extra-hippocampal neuronal network activity and, thereof, to impact the learning and memory function. The master biological clock located in the suprachiasmatic nucleus exerts both neural and endocrine (through the HHG axis-generating E2) controls upon brain neurons sub-serving the hippocampus-dependent learning and memory function. 5HT: 5-hydroxy-tryptamine (serotonin), ACh: acetylcholine, DA: dopamine, E2: estradiol, GABA: γ-aminobutyric acid, Glu: glutamate, HHG axis: hypothalamo-hypophyseal-gonadal axis, NE: norepinephrine.

All along the present chapter, we provided data that have reported some highly interesting effects reflecting both the neuroprotective and the neurogenic properties of isoflavones. However, the respective studies should be redesigned with lower concentrations of isoflavones and under dietary/supplemental configurations, as well, in order to get full acceptance as beneficial effects arising from nutritional isoflavones. Moreover, and regardless of the administration route/form of isoflavones, these substances did produce a variety of biological responses that have been shown to depend specifically on the sex, dose, age, animal model, hormonal status, exposure time, cell type and behavioral task. These occurrences make it evident that isoflavones, definitely, will not suite all the human beings nor every single health or aging-related condition. Consequently, we could not close this chapter without raising the question of the code of good practice, concerning the consumption of nutritional isoflavone-enriched products. Indeed, and to our knowledge, the only study that did address the question of isoflavone-mediated modulation of the activity of cerebral steroid-metabolizing enzymes under the dietary isoflavone intake status is the one previously published in isoflavone-rich diet fed rats and where the enzymatic activity of aromatase P450-CYP19 was measured in the preoptic/medio-basal hypothalamic regions. In that study, the lack of effect of isoflavones on the aromatase activity, in the single brain area examined to date, must underlies, rather than occludes, the need of investigating the impact of dietary/supplemental isoflavones on the acitivity of neurosteroidogenic enzymes in multiple brain regions including the hippocampal formation. More particularly, this impact should be evaluated in the contexts of both normal and pathophysiological brain aging, as well as in that of experimental E_2 withdrawal with and without estrogen replacement strategies. As a consequence of all these observations, it makes it evident that, beyond the tempting advertisement of isoflavone supplemental makers, the common sense of a safe use of isoflavones remains seeking the advice of a physician.

Dietary/supplemental isoflavones, obviously, displayed some beneficial effects on mood/anxiety aspects in both animal models and human subjects. The findings that isoflavones are efficient modulators of behaviorally/verbally expressed emotional states do agree with reports on the impact of estrogens on the mood function. Could it be possible that dietary/supplemental isoflavones induce some mild sedation-like effects that made some human subjects feel more relaxed/restrained and less tense (as they did express) and others, including animal subjects under isoflavone diets, better performers in behavioral cognitive tasks? Data that we did report in this chapter, such as the down-regulation of both cell excitability and $[Ca^{++}]_i$, would substantiate this suggestion. Once again, the corresponding studies remain to be redesigned with the nutritional constraints of isoflavone bio-availablity in mind.

The beneficial effects of dietary/supplemental isoflavones on mood-related cognition may be the iceberg tip of the impact of these molecules on the cerebral signalization involving neurotransmitter neuronal systems, namely GABAergic, glutamatergic, catecholaminergic, dopaminergic, serotoninergic and basal forebrain cholinergic neurons. More specifically, dietary/supplemental isoflavones did show interesting and beneficial leverage with both central nervous system cholinergic signalization and neuroprotection of basal forebrain neurons against age-triggered defects. Could it be possible that this same leverage with the modulation of cerebral cholinergic signalization is the one that did allow isoflavones to impact positively the expression of both the affective states and the hippocampus-linked mnemonic behaviors? If such positive potentials are really substantiated by future studies, then, sparing forebrain cholinergic neurons from age-triggering neurodegeneration may participate in rescuing the aging hippocampal formation and, thereof, in relieving age-related impairments of the memory function that are widespread among elderly men and women, and more particularly among those victims of neurodegenerative diseases.

References

Abdul-Ghani, MA; Valiante, TA;, Carlen, PL;, Pennefather, PS. Tyrosine kinase inhibitors enhance a Ca(2+)-activated K+ current (IAHP) and reduce IAHP suppression by a metabotropic glutamate receptor agonist in rat dentate granule neurones. *J. Physiol.* (1996). *496*(Pt 1):139-144.

Adams, MM; Shah, RA; Janssen, WG; Morrison, JH. Different modes of hippocampal plasticity in response to estrogen in young and aged female rats. *Proc. Natl. Acad. Sci. U. S. A.* (2001). *98*(14):8071-8076.

Adlercreutz, H; Markkanen, H; Watanabe, S. Plasma concentrations of phyto-oestrogens in Japanese men. *Lancet.* (1993). *342*(8881):1209-1210.

Agís-Balboa, RC; Pinna, G; Zhubi, A; Maloku, E; Veldic, M; Costa, E; Guidotti, A. Characterization of brain neurons that express enzymes mediating neurosteroid biosynthesis. *Proc. Natl. Acad. Sci. U. S. A.* (2006). *103*(39):14602-14607.

Albertazzi, P; Purdie, DW. Reprint of The nature and utility of the phytoestrogens: A review of the evidence. *Maturitas.* (2008). *61*(1-2):214-226.

Anderson, RL; Wolf, WJ. Compositional changes in trypsin inhibitors, phytic acid, saponins and isoflavones related to soybean processing. *J. Nutr.* (1995). *125*:581S-588S.

Aydin, M; Yilmaz, B; Alcin, E; Nedzvetsky, VS; Sahin, Z; Tuzcu, M. Effects of letrozole on hippocampal and cortical catecholaminergic neurotransmitter levels, neural cell adhesion molecule expression and spatial learning and memory in female rats. *Neuroscience.* (2008). *151*(1):186-194.

Azcoitia, I; Moreno, A; Carrero, P; Palacios, S; Garcia-Segura, LM. Neuroprotective effects of soy phytoestrogens in the rat brain. *Gynecol. Endocrinol.* (2006). *22*(2):63-69.

Bang, OY; Hong, HS; Kim, DH;, Kim, H; Boo, JH; Huh, K; Mook-Jung, I. Neuroprotective effect of genistein against beta amyloid-induced neurotoxicity. *Neurobiol. Dis.* (2004). *16*(1):21-28.

Barrett-Connor, E; Laughlin, GA. Endogenous and exogenous estrogen, cognitive function and dementia in postmenopausal women: Evidence from epidemiologic studies and clinical trials *Semin. Reprod. Med.* (2009). *27*(3): 275-282.

Bartlett, SE. Protein tyrosine kinase inhibitors synergize with nerve growth factor in embryonic chick sensory neuronal cell survival. *Neurosci Lett.* (1997). *227*(2):87-90.

Beattie, EC; Carroll, RC; Yu, X; Morishita, W; Yasuda, H; von Zastrow, M; Malenka, RC. Regulation of AMPA receptor endocytosis by a signaling mechanism shared with LTD. *Nat. Neurosci.* (2000). *3*(12):1291-1300.

Belcher, SM; Zsarnovszky, A. Estrogenic actions in the brain: Estrogen, phytoestrogens, and rapid intracellular signaling mechanisms. *J. Pharmacol. Exp. Ther.* (2001). *299*(2):408-414.

Bellino, FL; Wise, PM. Nonhuman primate models of menopause workshop. *Biol. Reprod.* (2003). *68*(1):10-18.

Bennetau-Pelissero, C; Arnal-Schnebelen, B; Lamothe, V; Sauvant, P; Sagne, JL; Verbruggen, MA; Mathey, J; Lavialle, O. ELISAs as new methods to measure isoflavones in food and human fluids. *Food Chem.* (2003). *82*:645-658.

Bennetau-Pelissero, C; Latonnelle, K; Séqueira, A; Lamothe, V. Phytœstrogens, possible endocrine disrupters from food. *Analusis.* (2000). *28*(9):763-775.

Bennetts, HW; Underwood, EJ; Shier, FL. A specific breeding problem of sheep on subterranean clover pastures in Western Australia. *Aust. Vet. J.* (1946). *22*:2-12.

Benton, AL. Differential behavioral effects in frontal lobe disease. *Neuropsychologia.* (1968). *6*:53-60.

Billard, JM. Aging, hippocampal synaptic activity and magnesium. *Magnes Res.* (2006). *19*(3):199-215.

Blomquist, CH; Lima, PH; Hotchkiss, JR. Inhibition of 3alpha-hydroxysteroid dehydrogenase (3alpha-HSD) activity of human lung microsomes by genistein, daidzein, coumestrol and C(18)-, C(19)- and C(21)-hydroxysteroids and ketosteroids. *Steroids.* (2005). *70*(8):507-514.

Boon, WC; Diepstraten, J; van der Burg, J; Jones, ME; Simpson, ER; van den Buuse, M. Hippocampal NMDA receptor subunit expression and watermaze learning in estrogen deficient female mice. *Brain Res. Mol. Brain Res.* (2005). *140*(1-2):127-132.

Boulware, MI; Mermelstein, PG. Membrane estrogen receptors activate metabotropic glutamate receptors to influence nervous system physiology. *Steroids.* (2009). *74*(7):608-613.

Braden, AWH; Hart, NK; Lamberton, JA. The œstrogenic activity and metabolism of certain isoflavones in sheep. *Aust. J. Agric. Res.* (1967). *18*:335-348.

Brailoiu, E; Dun, SL; Brailoiu, GC; Mizuo, K; Sklar, LA; Oprea, TI; Prossnitz, ER; Dun, NJ. Distribution and characterization of estrogen receptor G protein-coupled receptor 30 in the rat central nervous system. *J. Endocrinol.* (2007). *193*(2):311-321.

Brake, WG; Alves, SE; Dunlop, JC; Lee, SJ; Bulloch, K; Allen, PB; Greengard, P; McEwen, BS. Novel target sites for estrogen action in the dorsal hippocampus: An examination of synaptic proteins. *Endocrinology.* (2001). *142*(3):1284-1289.

Brann, DW: Dhandapani, K; Wakade, C; Mahesh, VB; Khan, MM. Neurotrophic and neuroprotective actions of estrogen: Basic mechanisms and clinical implications. *Steroids.* (2007). *72*(5):381-405.

Brann, DW; Zhang, QG;, Wang, RM; Mahesh, VB; Vadlamudi, RK. PELP1-- a novel estrogen receptor-interacting protein. *Mol. Cell Endocrinol.* (2008). *290*(1-2):2-7.

Bu, L; Lephart, ED. AVPV neurons containing estrogen receptor-beta in adult male rats are influenced by soy isoflavones. *BMC Neurosci.* (2007). 8:13.

Carlson, S; Peng, N; Prasain, JK; Wyss, JM. Effects of botanical dietary supplements on cardiovascular, cognitive, and metabolic function in males and females. *Gend. Med.* (2008). 5 (Suppl. A):76-90.

Casey, M; Maguire, C; Kelly, A; Gooney, MA; Lynch, MA. Analysis of the presynaptic signaling mechanisms underlying the inhibition of LTP in rat dentate gyrus by the tyrosine kinase inhibitor, genistein. *Hippocampus.* (2002). *12*(3):377-385.

Casini, ML; Marelli, G; Papaleo, E; Ferrari, A; D'Ambrosio, F; Unfer, V. Psychological assessment of the effects of treatment with phytoestrogens on postmenopausal women: A randomized, double-blind, crossover, placebo-controlled study. *Fertil. Steril.* (2006). *85*(4):972-978.

CDC: The Centers for Disease Control and Prevention. Public health and aging: Trends in aging--United States and worldwide. *JAMA*. (2003). *289*(11):1371-1373.

Charpantier, E; Wiesner, A; Huh, KH; Ogier, R; Hoda, JC; Allaman, G; Raggenbass, M; Feuerbach, D; Bertrand, D; Fuhrer, C. Alpha7 neuronal nicotinic acetylcholine receptors are negatively regulated by tyrosine phosphorylation and Src-family kinases. *J. Neurosci.* (2005). *25*(43):9836-9849.

Cho, CH; Song, W; Leitzell, K; Teo, E; Meleth, AD; Quick, MW; Lester, RA. Rapid upregulation of alpha7 nicotinic acetylcholine receptors by tyrosine dephosphorylation. *J. Neurosci.* (2005). *25*(14):3712-3723.

Choi, JM; Romeo, RD; Brake, WG; Bethea, CL; Rosenwaks, Z; McEwen, BS. Estradiol increases pre- and post-synaptic proteins in the CA1 region of the hippocampus in female rhesus macaques (*Macaca mulatta*). *Endocrinology*. (2003). *144*(11):4734-4738.

Christensen, K; Doblhammer, G; Rau, R; Vaupel, JW. Aging populations: The challenges ahead. *Lancet*. (2009). *374*(9696):1196-1208.

Chung, WC; Pak, TR; Suzuki, S; Pouliot, WA; Andersen, ME; Handa, RJ. Detection and localization of an estrogen receptor beta splice variant protein (ERbeta2) in the adult female rat forebrain and midbrain regions. *J. Comp. Neurol.* (2007). *505*(3):249-267.

Clarkson, TB; Anthony, MS; Hughes, CL Jr. Estrogenic soybean isoflavones and chronic disease Risks and benefits. *Trends Endocrinol. Metab.* (1995). *6*(1):11-16.

Cornwell, T; Cohick, W; Raskin, I. Dietary phytoestrogens and health. *Phytochemistry*. (2004). *65*(8):995-1016.

Cowansage, KK; Ledoux, JE; Monfils, MH. Brain-derived neurotrophic factor: A dynamic gatekeeper of neural plasticity. *Curr. Mol. Pharmacol.* (2009). Dec. 23. [Epub ahead of print].

Coward, L; Smith, M; Kirk, M; Barnes, S. Chemical modification of isoflavones in soyfoods during cooking and processing. *Am. J. Clin. Nutr.* (1998). *68*(Suppl. 6):1486S-1491S.

Crews, DE. Senescence, aging, and disease. *J. Physiol. Anthropol.* (2007). *26*(3):365-372.

Curran, HV; Schifano, F; Lader, M. Models of memory dysfunction? Acomparison of the effects of scopolamine and lorazepam on memory, psychomotor performance and mood. *Psychopharmacology*. (1991). 103:83-90.

Curtis, L; Buisson, B; Bertrand, S; Bertrand, D. Potentiation of human alpha4beta2 neuronal nicotinic acetylcholine receptor by estradiol. *Mol. Pharmacol.* (2002). *61*(1):127-135.

Damaj, MI. Influence of gender and sex hormones on nicotine acute pharmacological effects in mice. *J. Pharmacol. Exp. Ther.* (2001). *296*(1):132-140.

Daniel, JM; Hulst, JL; Berbling, JL. Estradiol replacement enhances working memory in middle-aged rats when initiated immediately after ovariectomy but not after a long-term period of ovarian hormone deprivation. *Endocrinology.* (2006). *147*(1):607-614.

Davis, C; Bradshaw, CM; Szabadi, E. The doors and People Memory Test: Validation of norms and some new correction formulae. *Br. J. Clin. Psychol.* (1999). 38(pt. 3):305-314.

De Lacalle, S. Estrogen effects on neuronal morphology. *Endocrine.* (2006). *29*(2):185-190.

De Lacalle, S. Dehydrogenase and aromatase in human granulosa-luteal cells. *J. Steroid Biochem. Mol. Biol.* (2005). *96*(3-4):279-286.

Downs, JL; Urbanski, HF. Neuroendocrine changes in the aging reproductive axis of female rhesus macaques (*Macaca mulatta*). *Biol. Reprod.* (2006). *75*(4):539-546.

Downs, JL; Wise, PM. The role of the brain in female reproductive aging. *Mol. Cell Endocrinol.* (2009). *299*(1):32-38.

Driscoll, I; Hamilton, DA; Petropoulos, H; Yeo, RA; Brooks, WM; Baumgartner, RN; Sutherland, RJ. The aging hippocampus: Cognitive, biochemical and structural findings. *Cereb. Cortex.* (2003). *13*(12):1344-1351.

Duffy, R; Wiseman, H; File, SE. Improved cognitive function in postmenopausal women after 12 weeks of consumption of a soya extract containing isoflavones. *Pharmacol. Biochem. Behav.* (2003). *75*(3):721-729.

Dumas, J; Hancur-Bucci, C; Naylor, M; Sites, C; Newhouse, P. Estradiol interacts with the cholinergic system to affect verbal memory in postmenopausal women: Evidence for the critical period hypothesis. *Horm. Behav.* (2008). *53*(1):159-169.

Edmunds, KM; Holloway, AC; Crankshaw, DJ; Agarwal, SK; Foster, WG. The effects of dietary phytoestrogens on aromatase activity in human endometrial stromal cells. *Reprod. Nutr. Dev.* (2005). *45*(6):709-720.

Favit, A; Fiore, L; Nicoletti, F; Canonico, PL. Estrogen modulates stimulation of inositol phospholipid hydrolysis by norepinephrine in rat brain slices. *Brain Res.* (1991). *555*(1):65-69

Fernandez, SM; Lewis, MC; Pechenino, AS; Harburger, LL; Orr, PT; Gresack, JE; Schafe, GE; Frick, KM. Estradiol-induced enhancement of object memory consolidation involves hippocampal extracellular signal-regulated kinase activation and membrane-bound estrogen receptors. *J. Neurosci.* (2008). Aug. 27; *28*(35):8660-8667.

Ferrari, CK. Functional foods, herbs and nutraceuticals: Towards biochemical mechanisms of healthy aging. *Biogerontology.* (2004). *5*(5):275-289.

Ferrucci, L; Giallauria, F; Guralnik, JM. Epidemiology of aging. *Radiol. Clin. North Am.* (2008). *46*(4):643-652, v.

Filardo, E; Quinn, J; Pang, Y; Graeber, C; Shaw, S; Dong, J; Thomas, P. Activation of the novel estrogen receptor G protein-coupled receptor 30 (GPR30) at the plasma membrane. *Endocrinology.* (2007). *148*(7):3236-3245.

File, SE; Hartley, DE; Alom, N; Rattray, M. Soya phytoestrogens change cortical and hippocampal expression of BDNF mRNA in male rats. *Neurosci. Lett.* (2003). *338*(2):135-138.

File, SE; Jarrett, N; Fluck, E; Duffy, R; Casey, K; Wiseman, H. Eating soya improves human memory. *Psychopharmacology.* (2001). *157*(4):430-436.

Findlay, JK; Buckmaster, JM; Chamley, WA; Cumming, IA; Hearnshaw, H; Goding, JR. Release of luteinising hormone by œstradiol 17β and a gonadotrophin-releasing hormone in ewes affected with clover disease. *Neuroendocrinology.* (1973). *11*:57-66.

Fitzpatrick, LA. Soy isoflavones: Hope or hype? *Maturitas.* (2003). *44* (Suppl. 1):S-21-S29.

Foster, TC. Involvement of hippocampal synaptic plasticity in age-related memory decline. *Brain Res. Brain Res. Rev.* (1999). *30*(3):236-249. Review. Pub. Med. PMID: 10567726.

Fournier, LR; Ryan-Borchers, TA; Robison, LM; Wiediger, M; Park, JS; Chew, BP; McGuire, MK; Sclar, DA; Skaer, TL; Beerman, KA. The effects of soy milk and isoflavone supplements on cognitive performance in healthy, postmenopausal women. *J. Nutr. Health Aging.* (2007). *11*(2):155-164.

Foy, MR; Baudry, M; Diaz-Brinton, R; Thompson, RF. Estrogen and hippocampal plasticity in rodent models. *J. Alzheimers Dis.* (2008). *15*(4):589-603.

Foy, MR; Xu, J; Xie, X; Brinton, RD; Thompson, RF; Berger, TW. 17beta-estradiol enhances NMDA receptor-mediated EPSPs and long-term potentiation. *J. Neurophysiol.* (1999). *81*(2):925-929.

Frick, KM; Fernandez, SM; Bulinski, SC. Estrogen replacement improves spatial reference memory and increases hippocampal synaptophysin in aged female mice. *Neuroscience.* (2002).*115*(2):547-558.

Frick, KM. Estrogens and age-related memory decline in rodents: What have we learned and where do we go from here? *Horm. Behav.* (2009). *55*(1):2-23.

Frye, CA; Rhodes, ME; Dudek, B. Estradiol to aged female or male mice improves learning in inhibitory avoidance and water maze tasks. *Brain Res.* (2005).*1036*(1-2):101-108.

Frye, CA. Estrus-associated decrements in a water maze task are limited to acquisition. *Physiol. Behav.* (1995). *57*(1):5-14.

Garcia-Segura, LM; Azcoitia, I; DonCarlos, LL. Neuroprotection by estradiol. *Prog. Neurobiol.* (2001). *63*(1):29-60.

Garcia-Segura, LM. Aromatase in the brain: Not just for reproduction anymore. *J. Neuroendocrinol.* (2008). *20*(6):705-712.

Gazzaley, AH; Weiland, NG; McEwen, BS; Morrison, JH. Differential regulation of NMDAR1 mRNA and protein by estradiol in the rat hippocampus. *J. Neurosci.* (1996). *16*(21):6830-6838.

Geller, SE; Studee, L. Soy and red clover for mid-life and aging. *Climacteric.* (2006). *9*(4):245-263.

Genazzani, AR; Pluchino, N; Luisi, S; Luisi, M. Estrogen, cognition and female aging. *Hum. Reprod. Update.* (2007). *13*(2):175-187.

Giovannini, MG; Pazzagli, M; Malmberg-Aiello, P; Della Corte, L; Rakovska, AD; Cerbai, F; Casamenti, F; Pepeu, G. Inhibition of acetylcholine-induced activation of extracellular regulated protein kinase prevents the encoding of an inhibitory avoidance response in the rat. *Neuroscience.* (2005).*136*(1):15-32.

Giovannini, MG. The role of the extracellular signal-regulated kinase pathway in memory encoding. *Rev. Neurosci.* (2006).*17*(6):619-634.

Gleason, CE; Carlsson, CM; Barnet, JH; Meade, SA; Setchell, KD; Atwood, CS; Johnson, SC; Ries, ML; Asthana, S. A preliminary study of the safety, feasibility and cognitive efficacy of soy isoflavone supplements in older men and women. *Age Ageing.* (2009). *38*(1):86-93.

Golub, MS; Hogrefe, CE; Germann, SL; Tran, TT; Beard, JL; Crinella, FM; Lonnerdal, B. Neurobehavioral evaluation of rhesus monkey infants fed cow's milk formula, soy formula, or soy formula with added manganese. *Neurotoxicol. Teratol.* (2005). *27*(4):615-627.

Gruber, CJ; Gruber, DM; Gruber, IML; Wieser, F; Huber, JC. Anatomy of the estrogen response element. *Trends Endocrinol. Metab.* (2004). *15*(2):73-78.

Halbreich, U; Lumley, LA; Palter, S; Manning, C; Gengo, F; Joe, SH. Possible acceleration of age effects on cognition following menopause. *J. Psychiatr. Res.* (1995). *29*(3):153-163.

Hall, JE. Neuroendocrine changes with reproductive aging in women. *Semin. Reprod. Med.* (2007). *25*(5):344-351.

Hazell, GG; Yao, ST; Roper, JA; Prossnitz, ER; O'Carroll, AM; Lolait, SJ. Localization of GPR30, a novel G protein-coupled oestrogen receptor, suggests multiple functions in rodent brain and peripheral tissues. *J. Endocrinol.* (2009). *202*(2):223-236.

Hellström-Lindahl, E; Mousavi, M; Zhang, X; Ravid, R; Nordberg, A. Regional distribution of nicotinic receptor subunit mRNAs in human brain: comparison between Alzheimer and normal brain. *Brain Res. Mol. Brain Res.* (1999). *66*(1-2):94-103.

Hill, AJ. First occurrence of hippocampal spatial firing in a new environment. *Exp. Neurol.* (1978). *62*(2):282-297.

Ho, SC; Chan, AS; Ho, YP; So, EK; Sham, A; Zee, B; Woo, JL. Effects of soy isoflavone supplementation on cognitive function in Chinese postmenopausal women: A double-blind, randomized, controlled trial. *Menopause.* (2007). *14*(3 Pt 1):489-499.

Hogervorst, E; Sadjimim, T; Yesufu, A; Kreager, P; Rahardjo, TB. High tofu intake is associated with worse memory in elderly Indonesian men and women. *Dement. Geriatr. Cogn. Disord.* (2008). *26*(1):50-57.

Hojo, Y; Murakami, G; Mukai, H; Higo, S; Hatanaka, Y; Ogiue-Ikeda, M; Ishii, H; Kimoto, T; Kawato, S. Estrogen synthesis in the brain--role in synaptic plasticity and memory. *Mol. Cell Endocrinol.* (2008). *290*(1-2):31-43.

Hooper, L; Ryder, JJ; Kurzer, MS; Lampe, JW; Messina, MJ; Phipps, WR; Cassidy, A. Effects of soy protein and isoflavones on circulating hormone concentrations in pre- and post-menopausal women: A systematic review and meta-analysis. *Hum. Reprod. Update.* (2009). *15*(4):423-440.

Huang, MH; Luetters, C; Buckwalter, GJ; Seeman, TE; Gold, EB; Sternfeld, B; Greendale, GA. Dietary genistein intake and cognitive performance in a multiethnic cohort of midlife women. *Menopause.* (2006). *13*(4):621-630.

Hung, SP; Hsu, JR; Lo, CP; Huang, HJ; Wang, JP; Chen, ST. Genistein-induced neuronal differentiation is associated with activation of extracellular signal-regulated kinases and upregulation of p21 and N-cadherin. *J. Cell Biochem.* (2005). *96*(5):1061-1070.

Ish, H; Tsurugizawa, T; Ogiue-Ikeda, M; Asashima, M; Mukai, H; Murakami, G; Hojo, Y; Kimoto, T; Kawato, S. Local production of sex hormones and their modulation of hippocampal synaptic plasticity. *Neuroscientist.* (2007). *13*(4):323-334.

Ishunina, TA; Fischer, DF; Swaab, DF. Estrogen receptor alpha and its splice variants in the hippocampus in aging and Alzheimer's disease. *Neurobiol. Aging.* (2007). *28*(11):1670-1681.

Izumi, T; Piskula, MK; Osawa, S; Obata, A; Tobe, K; Saito, M; Kataoka, S; Kubota, Y; Kikuchi, M. Soy isoflavone aglycones are absorbed faster and in higher amounts than their glucosides in humans. *J. Nutr.* (2000). *130*(7):1695-1699.

Jefremov, V; Rakitin, A; Mahlapuu, R; Zilmer, K; Bogdanovic, N; Zilmer, M; Karelson, E. 17beta-Oestradiol stimulation of G-proteins in aged and Alzheimer's human brain: Comparison with phytoestrogens. *J. Neuroendocrinol.* (2008). *20*(5):587-596.

Kajta, M; Domin, H; Grynkiewicz, G; Lason, W. Genistein inhibits glutamate-induced apoptotic processes in primary neuronal cell cultures: An involvement of aryl hydrocarbon receptor and estrogen receptor/glycogen synthase kinase-3beta intracellular signaling pathway. *Neuroscience.* (2007). *145*(2):592-604.

Kalita, K; Szymczak, S; Kaczmarek, L. Non-nuclear estrogen receptor beta and alpha in the hippocampus of male and female rats. *Hippocampus.* (2005). *15*(3):404-412.

Kaplan, E; Goodglass, H; Weintraub, S. *The Boston Naming Test.* Philadelphia, Pa: Lea & Febiger. 1982.

Katzenellenbogen, JA; Katzenellebogen, BS; Tatee, T; Robertson, DW; Landvatter, SW. The chemistry of estrogens and antiestrogens: Relationships between structure, receptor binding ans biological activity. In: JA McLachlan (Ed.), *Estrogens in the Environment.* Elsevier Inc. Pub, (1980). 33-51.

Kelly, MJ; Levin, ER. Rapid actions of plasma membrane estrogen receptors. *Trends End. Metab.* (2001). *12*:152-156.

Kelly, MJ; Rønnekleiv, OK. Membrane-initiated estrogen signaling in hypothalamic neurons. *Molec. Cel. Endocrinol.* (2008). *290*:14-23.

Kelly, MJ; Wagner, EJ. Estrogen modulation of G-protein-coupled receptors. *Trends End Metab.* (1999). *10*:369-374.

Kim, JS; Kim, HY; Kim, JH; Shin, HK; Lee, SH; Lee, YS; Son, H. Enhancement of rat hippocampal long-term potentiation by 17 beta-estradiol involves mitogen-activated protein kinase-dependent and -independent components. *Neurosci. Lett.* (2002). *332*(1):65-69.

Kitamura, N; Araya, R; Kudoh, M; Kishida, H; Kimura, T; Murayama, M; Takashima, A; Sakamaki, Y; Hashikawa, T; Ito, S; Ohtsuki, S; Terasaki, T; Wess, J; Yamada, M. Beneficial effects of estrogen in a mouse model of cerebrovascular insufficiency. *PLoS One.* (2009). *4*(4):e5159.

Knaus, HG; Schwarzer, C; Koch, RO; Eberhart, A; Kaczorowski, GJ; Glossmann, H; Wunder, F; Pongs, O; Garcia, ML; Sperk, G. Distribution of high-conductance Ca(2+)-activated K+ channels in rat brain: Targeting to axons and nerve terminals. *J. Neurosci.* (1996). *16*(3):955-963.

Koss, DJ; Riedel, G; Platt, B. Intracellular Ca2+ stores modulate SOCCs and NMDA receptors via tyrosine kinases in rat hippocampal neurons. *Cell Calcium.* (2009). *46*(1):39-48.

Kreijkamp-Kaspers, S; Kok, L; Grobbee, DE; de Haan, EH; Aleman, A; Lampe, JW; vanderSchouw, YT. Effect of soy protein containing isoflavones on cognitive function, bone mineral density and plasma lipids in postmenopausal women: A randomized controlled trial. *JAMA.* (2004). *292*(1):65-74.

Kreijkamp-Kaspers, S; Kok, L; Grobbee, DE; de Haan, EH; Aleman, A; vanderSchouw, YT. Dietary phytoestrogen intake and cognitive function in older women. *J. Gerontol. A Biol. Sci. Med. Sci.* (2007). *62*(5):556-562.

Kretz, O; Fester, L; Wehrenberg, U; Zhou, L; Brauckmann, S; Zhao, S; Prange-Kiel, J; Naumann, T; Jarry, H; Frotscher, M; Rune, GM. Hippocampal synapses depend on hippocampal estrogen synthesis. *J. Neurosci.* (2004). *24*(26):5913-5921.

Kritz-Silverstein, D; Von Mühlen, D; Barrett-Connor, E; Bressel, MA. Isoflavones and cognitive function in older women: The SOy and Postmenopausal Health In Aging (SOPHIA) Study. *Menopause.* (2003). *10*(3):196-202.

Kuiper, GC; Carlsson, B; Grandien, K; Enmark, E; Haggblad, J; Nilsson, S; Gusfasson, JA. Comparison of the ligand binding specificity and transcript tissue distribution of estrogen receptors alpha and beta. *Endocrinology.* (1997). *138*(3):863-870.

Kuiper, GG; Lemmen, JG; Carlsson, B; Corton, JC; Safe, SH; van der Saag, PT; van der Burg, B; Gustafsson, JA. Interaction of estrogenic chemicals and phytoestrogens with estrogen receptor beta. *Endocrinology.* (1998). *139*(10):4252-4263.

Kumar, A; Foster, TC. 17beta-estradiol benzoate decreases the AHP amplitude in CA1 pyramidal neurons. *J. Neurophysiol.* (2002). *88*(2):621-626.

Lacey, M; Bohday, J; Fonseka, SM; Ullah, AI; Whitehead, SA. Dose-response effects of phytoestrogens on the activity and expression of 3beta-hydroxysteroid dehydrogenase and aromatase in human granulosa-luteal cells. *J. Steroid Biochem. Mol. Biol.* (2005). *96*(3-4):279-286.

Lacreuse, A. Effects of ovarian hormones on cognitive function in nonhuman primates. *Neuroscience.* (2006). *138*(3):859-867.

Lâm, TT; Leranth, C. Role of the medial septum diagonal band of Broca cholinergic neurons in oestrogen-induced spine synapse formation on hippocampal CA1 pyramidal cells of female rats. *Eur. J. Neurosci.* (2003). *17*(10):1997-2005.

Lee, YB; Lee, HJ; Sohn, HS. Soy isoflavones and cognitive function. *J. Nutr. Biochem.* (2005). *16*(11):641-649.

Lee, YB; Lee, HJ; Won, MH; Hwang, IK; Kang, TC; Lee, JY; Nam, SY; Kim, KS; Kim, E; Cheon, SH; Sohn, HS. Soy isoflavones improve spatial delayed matching-to-place performance and reduce cholinergic neuron loss in elderly male rats. *J. Nutr.* (2004). *134*(7):1827-1831.

Lee, YB; Lee, KH; Sohn, HS; Lee, SJ; Cho, KH; Kang, IJ; Kim, DW; Shin, YK; Pai, T; Hwang, IK; Won, MH. Effects of soy phytoestrogens on reference memory and neuronal cholinergic enzymes in ovariectomized rats. *J. Med. Food.* (2009). *12*(1):64-70.

Lee, SJ; Campomanes, CR; Sikat, PT; Greenfield, AT; Allen, PB; McEwen, BS. Estrogen induces phosphorylation of cyclic AMP response element binding (pCREB) in primary hippocampal cells in a time-dependent manner. *Neuroscience.* (2004).*124*(3):549-560.

Lephart, ED; Adlercreutz, H; Lund, TD. Dietary soy phytoestrogen effects on brain structure and aromatase in Long-Evans rats. *Neuroreport.* (2001). *12*(16):3451-3455.

Lephart, ED; Setchell, KD; Handa, RJ; Lund, TD. Behavioral effects of endocrine-disrupting substances: Phytoestrogens. *ILAR J.* (2004). *45*(4):443-454.

Lephart, ED; Setchell, KD; Lund, TD. Phytoestrogens: Hormonal action and brain plasticity. *Brain Res. Bull.* (2005). *65*(3):193-198.

Lephart, ED; Thompson, JM; Setchell, KD; Adlercreutz, H; Weber, KS. Phytoestrogens decrease brain calcium-binding proteins but do not alter hypothalamic androgen metabolizing enzymes in adult male rats. *Brain Res.* (2000). *859*(1):123-131.

Levin, ER. Plasma membrane estrogen receptors. *Trends Endocrinol Metab.* (2009). *20*(10):477-482.

Levitzki, A; Mishani, E. Tyrphostins and other tyrosine kinase inhibitors. *Annu. Rev. Biochem.* (2006). *75*:93-109.

Lezak, MD. *Neuropsychological Assessment.* New York: Oxford University Press. (1995).

Linford, NJ; Yang, Y; Cook, DG; Dorsa, DM. Neuronal apoptosis resulting from high doses of the isoflavone genistein: Role for calcium and p42/44 mitogen-activated protein kinase. *J. Pharmacol. Exp. Ther.* (2001). *299*(1):67-75.

Liu, F; Day, M; Muñiz, LC; Bitran, D; Arias, R; Revilla-Sanchez, R; Grauer, S; Zhang, G; Kelley, C; Pulito, V; Sung, A; Mervis, RF; Navarra, R; Hirst, WD; Reinhart, PH; Marquis, KL; Moss, SJ; Pangalos, MN; Brandon, NJ. Activation of estrogen receptor-beta regulates hippocampal synaptic plasticity and improves memory. *Nat. Neurosci.* (2008). *11*(3):334-343.

Liu, YQ ; Xin, TR ; Lü, XY ; Ji, Q ; Jin, Y ; Yang, HD. Memory performance of hypercholesterolemic mice in response to treatment with soy isflavones. *Neurosci. Res.* (2007). *57*(4):544-549.

Lu, WY; Xiong, ZG; Lei, S; Orser, BA; Dudek, E; Browning, MD; MacDonald, JF. G-protein-coupled receptors act via protein kinase C and Src to regulate NMDA receptors. *Nat. Neurosci.* (1999). *2*(4):331-338.

Luine, V; Attalla, S; Mohan, G; Costa, A; Frankfurt, M. Dietary phytoestrogens enhance spatial memory and spine density in the hippocampus and prefrontal cortex of ovariectomized rats. *Brain Res. Rev.* (2006). *1126*(1):183-187.

Lund, TD; West, TW; Tian, LY; Bu, LH; Simmons, DL; Setchell, KD; Adlercreutz, H; Lephart, ED. Visual spatial memory is enhanced in female rats (but inhibited in males) by dietary soy phytoestrogens. *BMC Neurosci.* (2001). 2:20.

Lund, TD; Lephart, ED. Dietary soy phytoestrogens produce anxiolytic effects in the elevated plus-maze. *Brain Res.* (2001a). *913*(2):180-184.

Lund, TD; Lephart, ED. Manipulation of prenatal hormones and dietary phytoestrogens during adulthood alter the sexually dimorphic expression of visual spatial memory. *BMC Neurosci.* (2001b). 2:21.

Lutz, W; Qiang, R. Determinants of human population growth. *Philos Trans R Soc Lond B Biol Sci.* (2002). *357*(1425):1197-1210.

Maggiolini, M; Picard, D. The unfolding stories of GPR30, a new membrane-bound estrogen receptor. *J. Endocrinol.* (2010). *204*(2):105-114.

Maggiolini, M; Vivacqua, A; Fasanella, G; Recchia, AG; Sisci, D; Pezzi, V; Montanaro, D; Musti, AM; Picard, D; Andò, S. The G protein-coupled receptor GPR30 mediates c-fos up-regulation by 17beta-estradiol and phytoestrogens in breast cancer cells. *J. Biol. Chem.* (2004). *279*(26):27008-27016.

Maki, PM;, Dumas, J. Mechanisms of action of estrogen in the brain: Insights from human neuroimaging and psychopharmacologic studies. *Semin. Reprod. Med.* (2009). *27*(3):250-259.

Maki, PM;, Rubin, LH; Fornelli, D; Drogos, L; Banuvar, S; Shulman, LP; Geller, SE. Effects of botanicals and combined hormone therapy on cognition in postmenopausal women. *Menopause.* (2009). *16*(6):1167-1177.

Marrian, GF; Haselwood, GAD. Equol a new inactive phenol isolated from the ketohydroxy-œstrin fraction of mares' urine. *Biochem. J.* (1932). *26*:1227-1232.

Mathey, J; Lamothe, V; Coxam, V; Potier, M; Sauvant, P; Bennetau-Pelissero, C. Concentrations of isoflavones in plasma and urine of post-menopausal women chronically ingesting high quantities of soy isoflavones. *J. Pharm. Biomed. Anal.* (2006). *41*:957-965.

Mathey, J; Mardon, J; Fokialakis, N; Puel, C; Kati-Coulibaly, S; Mitakou, S; Bennetau-Pelissero, C; Lamothe, V; Davicco, MJ; Lebecque, P; Horcajada, MN; Coxam, V. Modulation of soy isoflavones bioavailability and subsequent effects on bone health in ovariectomized rats: The case for equol. *Osteoporos Int.* (2007). *18*(5):671-679.

Matsuda, K; Sakamoto, H; Mori, H; Hosokawa, K; Kawamura, A; Itose, M; Nishi, M; Prossnitz, ER; Kawata, M. Expression and intracellular distribution of the G protein-coupled receptor 30 in rat hippocampal formation. *Neurosci. Lett.* (2008). *441*(1):94-99.

McEwen, B. Estrogen actions throughout the brain. *Recent Prog. Horm. Res.* (2002). *57*:357-384.

McEwen, BS; Alves, SE; Bulloch, K; Weiland, NG. Ovarian steroids and the brain: Implications for cognition and aging. *Neurology.* (1997). *48*(5) Suppl. 7):S8-15.

McEwen, BS; Milner, TA. Hippocampal formation: Shedding light on the influence of sex and stress on the brain. *Brain Res. Rev.* (2007). *55*(2):343-355.

McNaughton, BL; Barnes, CA; O'Keefe, J. The contributions of position, direction, and velocity to single unit activity in the hippocampus of freely-moving rats. *Exp. Brain Res.* (1983). *52*(1):41-49.

Mehra, RD; Sharma, K; Nyakas, C; Vij, U. Estrogen receptor alpha and beta immunoreactive neurons in normal adult and aged female rat hippocampus: A qualitative and quantitative study. *Brain Res.* (2005). *1056*(1):22-35.

Mendelson, CR; Jiang, B; Shelton, JM; Richardson, JA; Hinshelwood, MM. Transcriptional regulation of aromatase in placenta and ovary. *J. Steroid Biochem. Mol. Biol.* (2005). *95*(1-5):25-33.

Messinis, IE. Ovarian feedback, mechanism of action and possible clinical implications. *Hum. Reprod. Update.* (2006). *12*(5):557-571.

Mhyre, AJ; Dorsa, DM. Estrogen activates rapid signaling in the brain: Role of estrogen receptor alpha and estrogen receptor beta in neurons and glia. *Neuroscience.* (2006). *138*(3):851-858.

Mielke, JG; Mealing, GA. Cellular distribution of the nicotinic acetylcholine receptor alpha7 subunit in rat hippocampus. *Neurosci. Res.* (2009). *65*(3):296-306.

Milner, TA; McEwen, BS; Hayashi, S; Li, CJ; Reagan, LP; Alves, SE. Ultrastructural evidence that hippocampal alpha estrogen receptors are located at extranuclear sites. *J. Comp. Neurol.* (2001). *429*(3):355-371.

Milner, TA; Ayoola, K; Drake, CT; Herrick, SP; Tabori, NE; McEwen, BS; Warrier, S; Alves, SE. Ultrastructural localization of estrogen receptor beta immunoreactivity in the rat hippocampal formation. *J. Comp. Neurol.* (2005). *491*(2):81-95.

Milner, TA; Lubbers, LS; Alves, SE; McEwen, BS. Nuclear and extranuclear estrogen binding sites in the rat forebrain and autonomic medullary areas. *Endocrinology.* (2008). *149*(7):3306-3312.

Mitra, SW; Hoskin, E; Yudkovitz, J; Pear, L; Wilkinson, HA; Hayashi, S; Pfaff, DW; Ogawa, S; Rohrer, SP; Schaeffer, JM; McEwen, BS; Alves, SE. Immunolocalization of estrogen receptor beta in the mouse brain: Comparison with estrogen receptor alpha. *Endocrinology.* (2003). *144*(5):2055-2067. Erratum in: *Endocrinology.* (2003). *144*(7):2844.

Molokanova, E; Savchenko, A; Kramer, RH. Interactions of cyclic nucleotide-gated channel subunits and protein tyrosine kinase probed with genistein. *J. Gen. Physiol.* (2000). *115*(6):685-696.

Molokanova, E; Savchenko, A; Kramer, RH. Noncatalytic inhibition of cyclic nucleotide-gated channels by tyrosine kinase induced by genistein. *J. Gen. Physiol.* (1999). *113*(1):45-56.

Montague, D; Weickert, CS; Tomaskovic-Crook, E; Rothmond, DA; Kleinman, JE; Rubinow, DR. Oestrogen receptor alpha localization in the prefrontal cortex of three mammalian species. *J. Neuroendocrinol.* (2008). *20*(7):893-903.

Monteiro, SC; de Mattos, CB; Ben, J; Netto, CA; Wyse, AT. Ovariectomy impairs spatial memory: Prevention and reversal by a soy isoflavone diet. *Metab. Brain Dis*. (2008). *23*(3):243-253.

Moutsatsou, P. The spectrum of phytoestrogens in nature: Our knowledge is expanding. *Hormones.* (Athens). (2007). *6*(3):173-193.

Nomura, M; Akama, KT; Alves, SE; Korach, KS; Gustafsson, JA; Pfaff, DW; Ogawa, S. Differential distribution of estrogen receptor (ER)-alpha and ER-beta in the midbrain raphe nuclei and periaqueductal gray in male mouse: Predominant role of ER-beta in midbrain serotonergic systems. *Neuroscience.* (2005). *130*(2):445-456.

O'Dell, TJ; Kandel, ER; Grant, SG. Long-term potentiation in the hippocampus is blocked by tyrosine kinase inhibitors. *Nature.* (1991). *353*(6344):558-560.

Ogiue-Ikeda, M; Tanabe, N; Mukai, H; Hojo, Y; Murakami, G; Tsurugizawa, T; Takata, N; Kimoto, T; Kawato, S. Rapid modulation of synaptic plasticity by estrogens as well as endocrine disrupters in hippocampal neurons. *Brain Res. Rev.* (2008).*57*(2):363-375.

Ohlsson, A; Ullerås, E; Oskarsson, A. A biphasic effect of the fungicide prochloraz on aldosterone, but not cortisol, secretion in human adrenal H295R cells--underlying mechanisms. *Toxicol. Lett.* (2009). *191*(2-3):174-180.

Olton, DS; Branch, M; Best, PJ. Spatial correlates of hippocampal unit activity. *Exp. Neurol.* (1978). *58*(3):387-409.

Owen, AM; Downes, JJ; Sahakian, BJ; Polkey, CE; Robbins, TW. Planning and spatial working memory following frontal lobe lesions in man. *Neuropsychologia.* (1990). *28*:1021-1034.

Owen, AM; Roberts, AC; Polkey, CE; Sahakian, BJ; Robbins, TW. Extra-dimensional versus intra-dimensional set shifting performance following frontal lobe excisions, temporal lobe excisions or amygdalo-hyppocampectomy in man. *Neuropsychologia.* (1991). *29*:993-1006.

Owen, AM; Sahakian, BJ; Semple, J; Polkey, CE; Robbins, TW. Visuospatial short-term recognition memory and learning after temporal lobe excisions, frontal lobe excisions or amygdalo-hyppocampectomy in man. *Neuropsychologia.* (1995). *33*:1-24.

Pacheco-Otalora, LF; Hernandez, EF; Arshadmansab, MF; Francisco, S; Willis, M; Ermolinsky, B; Zarei, M; Knaus, HG; Garrido-Sanabria, ER. Down-regulation of BK channel expression in the pilocarpine model of temporal lobe epilepsy. *Brain Res.* (2008). *1200*:116-131.

Paillart, C; Carlier, E; Guedin, D; Dargent, B; Couraud, F. Direct block of voltage-sensitive sodium channels by genistein, a tyrosine kinase inhibitor. *J. Pharmacol. Exp. Ther.* (1997). *280*(2):521-526.

Pan, Y; Anthony, M; Clarkson, TB. Effect of estradiol and soy phytoestrogens on choline acetyltransferase and nerve growth factor mRNAs in the frontal cortex and hippocampus of female rats. *Pro. Soc. Exp. Biol. Med.* (1999a). *221*(2):118-125.

Pan, Y; Anthony, M; Clarkson, TB. Evidence for up-regulation of brain-derived neurotrophic factor mRNA by soy phytoestrogens in the frontal cortex of retired breeder female rats. *Neurosci. Lett.* (1999b).*12*:261(1-2):17-20.

Pan, Y; Anthony, M; Watson, S; Clarkson, TB. Soy phytoestrogens improve radial arm maze performance in ovariectomized retired breeder rats and do not attenuate benefits of 17beta-estradiol treatment. *Menopause.* (2000). *7*(4):230-235.

Pelissero, C; Lenczowski, M; Chinzi, D; Sumpter, JP; Fostier, A. Effects of flavonoids on aromatase activity, an *in vitro* study. *J. Steroid Biochem. Molec. Biol.* (1996). *57*(3/4):215-223.

Pelletier, G; Dupont, E; Simard, J; Luu-The, V; Bélanger, A; Labrie, F. Ontogeny and subcellular localization of 3 beta-hydroxysteroid dehydrogenase (3 beta-HSD) in the human and rat adrenal, ovary and testis. *J. Steroid Biochem. Mol. Biol.* (1992). *43*(5):451-467.

Phelps, EA; Hyder, F; Blamire, AM; Shulman, RG. fMRI of the prefrontal cortex during overt verbal fluency. *Neuroreport.* (1997). *8*:561-565.

Picherit, C; Coxam, V; Bennetau-Pelissero, C; Kati-Coulibaly, S; Davicco, MJ; Lebecque, P; Barlet, JP. Daidzein is more efficient than genistein in preventing ovariectomy-induced bone loss in rats. *J. Nutr.* (2000). *130*(7):1675-1681.

Pihlajamaki, SM; Tanila, H; Hanninen, T; Kononen, M; Laakso, M; Partanen, K; Soininen, H; Aronen, HJ. Verbal fluency activates the left medial temporal lobe: A functional magnetic resonance imaging study. *Ann. Neurol.* (2000). *47*:470-476.

Potier, B; Rovira, C. Protein tyrosine kinase inhibitors reduce high-voltage activating calcium currents in CA1 pyramidal neurones from rat hippocampal slices. *Brain Res.* (1999). *816*(2):587-597.

Prange-Kiel, J; Rune, GM; Leranth, C. Median raphe mediates estrogenic effects to the hippocampus in female rats. *Eur. J. Neurosci.* (2004). *19*(2):309-317.

Prange-Kiel, J; Rune, GM. Direct and indirect effects of estrogen on rat hippocampus. *Neuroscience.* (2006). *138*(3):765-772.

Prange-Kiel, J;, Fester, L; Zhou, L; Lauke, H; Carrétero, J; Rune, GM. Inhibition of hippocampal estrogen synthesis causes region-specific downregulation of synaptic protein expression in hippocampal neurons. *Hippocampus.* (2006). *16*(5):464-471.

Prossnitz, ER; Arterburn, JB; Sklar, LA. GPR30: A G protein-coupled receptor for estrogen. *Mol. Cell Endocrinol.* (2007). 265-266:138-142.

Qiu, J; Bosch, MA; Jamali, K; Xue, C; Kelly, MJ; Rønnekleiv, OK. Estrogen upregulates T-type calcium channels in the hypothalamus and pituitary. *J. Neurosci.* (2006). *26*(43):11072-11082.

Ramsey, MM; Adams, MM; Ariwodola, OJ; Sonntag, WE; Weiner, JL. Functional characterization of des-IGF-1 action at excitatory synapses in the CA1 region of rat hippocampus. *J. Neurophysiol.* (2005). *94*(1):247-254.

Reitan, RK; Wolfson, D. *The Halestead-Reitan Neuropsychological Test Battery.* Tucson, Arizonia: Neuropsychological Press. (1985).

Rice, S; Mason, HD; Whitehead, SA. Phytoestrogens and their low dose combinations inhibit mRNA expression and activity of aromatase in human granulosa-luteal cells. *J. Steroid Biochem. Mol. Biol.* (2006). *101*(4-5):216-225.

Rogalski, SL; Appleyard, SM; Pattillo, A; Terman, GW; Chavkin, C. TrkB activation by brain-derived neurotrophic factor inhibits the G protein-gated inward rectifier Kir3 by tyrosine phosphorylation of the channel. *J. Biol. Chem.* (2000). *275*(33):25082-25088.

Rowland, I; Faughnan, M; Hoey, L; Wähälä, K; Williamson, G; Cassidy, A. Bioavailability of phyto-oestrogens. *Br. J. Nutr.* (2003). *89* (Suppl 1):S45-58.

Rowland, LM; Astur, RS; Jung, RE; Bustillo, JR; Lauriello, J; Yeo, RA. Selective cognitive impairments associated with NMDA receptor blockade in humans. *Neuropsychopharmacology.* (2005). *30*(3):633-639.

Rune, GM; Frotscher, M. Neurosteroid synthesis in the hippocampus: Role in synaptic plasticity. *Neurosciences.* (2005). *136*(3):833-842.

Rune, GM; Wehrenberg, U; Prange-Kiel, J; Zhou, L; Adelmann, G; Frotscher, M. Estrogen up-regulates estrogen receptor alpha and synaptophysin in slice cultures of rat hippocampus. *Neuroscience.* (2002). *113*(1):167-175.

Sakamoto, H; Matsuda, K; Hosokawa, K; Nishi, M; Morris, JF; Prossnitz, ER; Kawata, M. Expression of G protein-coupled receptor-30, a G protein-coupled membrane estrogen receptor, in oxytocin neurons of the rat paraventricular and supraoptic nuclei. *Endocrinology.* (2007). *148*(12):5842-5850.

Sarkar, SN; Huang, RQ; Logan, SM; Yi, KD; Dillon, GH; Simpkins, JW. Estrogens directly potentiate neuronal L-type Ca2+ channels. *Proc. Natl. Acad. Sci. U.S.A.* (2008). *105*(39):15148-15153.

Scallet, AC; Wofford, M; Meredith, JC; Allaben, WT; Ferguson, SA. Dietary exposure to genistein increases vasopressin but does not alter beta-endorphin in the rat hypothalamus. *Toxicol. Sci.* (2003). *72*(2):296-300.

Scharfman, HE;, MacLusky, NJ. Estrogen and brain-derived neurotrophic factor (BDNF) in hippocampus: Complexity of steroid hormone-growth factor interactions in the adult CNS. *Front Neuroendocrinol.* (2006). *27*(4):415-435.

Schlosser, R; Hutchinson, M; Joseffer, S; Rusinek, H; Saarimaki, A; Stevenson, J; Dewey, SL; Brodie, JD. Functional magnetic resonance imaging of human brain activity in a verbal fluency task. *J. Neurol. Neurosurg Psychiatry.* (1998). *64*:492-498.

Schreihofer, DA; Redmond, L. Soy phytoestrogens are neuroprotective against stroke-like injury *in vitro*. *Neuroscience.* (2009). *158*(2):602-609.

Schreihofer, DA. Transcriptional regulation by phytoestrogens in neuronal cell lines. *Mol. Cell Endocrinol.* (2005). *231*(1-2):13-22.

Setchell, KD; Brown, NM; Lydeking-Olsen, E. The clinical importance of the metabolite equol-a clue to the effectiveness of soy and its isoflavones. *J. Nutr.* (2002). *132*(12):3577-3584.

Setchell, KD. Soy isoflavones—benefits and risks from nature's selective estrogen receptor modulators (SERMs). *J. Am. Coll Nutr.* (2001). *20*(Suppl. 5):354S-362S.

Setchell, KDR; Cassidy, A. Dietary isoflavones: Biological effects and relevance to human health. *J. Nutr.* (1999). *129*:758S-767S.

Shankar, S; Teyler, TJ; Robbins, N. Aging differentially alters forms of long-term potentiation in rat hippocampal area CA1. *J. Neurophysiol.* (1998). *79*(1):334-341.

Sherwin, BB. Estrogen and cognitive aging in women. *Neuroscience.* (2006).138(3):1021-1026.

Sherwin, BB. The clinical relevance of the relationship between estrogen and cognition in women. *J. Steroid Biochem. Mol. Biol.* (2007). *106*(1-5):151-156.

Shingo, AS; Kito, S. Estrogen induces insulin-like growth factor-1 mRNA expression in the immortalized hippocampal cell: Determination by quantitative real-time polymerase chain reaction. *Neurochem. Res.* (2003). *28*(9):1379-1383.

Shrestha, LB. Population aging in developing countries. *Health Aff.* (Millwood). (2000). *19*(3):204-212.

Shughrue, PJ; Merchenthaler, I. Evidence for novel estrogen binding sites in the rat hippocampus. *Neuroscience.* (2000). *99*(4):605-612.

Smith, CC; McMahon, LL. Estradiol-induced increase in the magnitude of long-term potentiation is prevented by blocking NR2B-containing receptors. *J. Neurosci.* (2006). *26*(33):8517-8522.

Smith, CC; McMahon, LL. Estrogen-induced increase in the magnitude of long-term potentiation occurs only when the ratio of NMDA transmission to AMPA transmission is increased. *J. Neurosci.* (2005). *25*(34):7780-7791.

Smith, CC; Vedder, LC; McMahon, LL. Estradiol and the relationship between dendritic spines, NR2B containing NMDA receptors, and the magnitude of long-term potentiation at hippocampal CA3-CA1 synapses. *Psychoneuroendocrinology.* (2009). *34* (Suppl. 1):S130-142.

Smith, MJ; Jennes, L. Neural signals that regulate GnRH neurones directly during the oestrous cycle. *Reproduction.* (2001). *122*(1):1-10.

Snyder, MA; Cooke, BM; Woolley, CS. Estradiol potentiation of NR2B-dependent EPSCs is not due to changes in NR2B protein expression or phosphorylation. *Hippocampus.* (2010). [Epub ahead of print].

Somponpun, S; Sladek, CD. Role of estrogen receptor-beta in regulation of vasopressin and oxytocin release *in vitro. Endocrinology.* (2002). *143*(8):2899-2904.

Son, JH; Winzer-Serhan, UH. Expression of neuronal nicotinic acetylcholine receptor subunit mRNAs in rat hippocampal GABAergic interneurons. *J. Comp. Neurol.* (2008). *511*(2):286-299.

Spencer, JL; Waters, EM; Romeo, RD; Wood, GE; Milner, TA; McEwen, BS. Uncovering the mechanisms of estrogen effects on hippocampal function. *Front Neuroendocrinol.* (2008). *29*(2):219-237.

Spreen, O; Strauss, E. *A compendium of neuropsychological tests: Administrarion norms and commentary.* Oxford University Press, Oxford. (1991).

Takanami, K; Sakamoto, H; Matsuda, K; Hosokawa, K; Nishi, M; Prossnitz, ER; Kawata, M. Expression of G protein-coupled receptor 30 in the spinal somatosensory system. *Brain Res.* (2010). *1310*:17-28.

Talboom, JS; Williams, BJ; Baxley, ER; West, SG; Bimonte-Nelson, HA. Higher levels of estradiol replacement correlate with better spatial memory in surgically menopausal young and middle-aged rats. *Neurobiol. Learn Mem.* (2008). *90*(1):155-163.

Thijssen, JH; Daroszewski, J; Milewicz, A; Blankenstein, MA. Local aromatase activity in human breast tissues. *J. Steroid Biochem. Mol. Biol.* (1993). *44*(4-6):577-582.

Thomas, P; Dong, J. Binding and activation of the seven-transmembrane estrogen receptor GPR30 by environmental estrogens: A potential novel mechanism of endocrine disruption. *J. Steroid Biochem. Mol. Biol.* (2006). *102*(1-5):175-179.

Thompson, MA; Lasley, BL; Rideout, BA; Kasman, LH. Characterization of the estrogenic properties of a nonsteroidal estrogen equol, extracted from the urine of pregnant macaques. *Biol. Reprod.* (1984). *31*:705-713.

Thorp, AA; Sinn, N; Buckley, JD; Coates, AM; Howe, PR. Soya isoflavone supplementation enhances spatial working memory in men. *Br. J. Nutr.* (2009). *1*:1-7.

Tohgi, H; Utsugisawa, K; Yoshimura, M; Nagane, Y; Mihara, M. Age-related changes in nicotinic acetylcholine receptor subunits alpha4 and beta2 messenger RNA expression in postmortem human frontal cortex and hippocampus. *Neurosci. Lett.* (1998). *245*(3):139-142.

Usui, T. Pharmaceutical prospects of phytoestrogens. *Endocr. J.* (2006). *53*(1):7-20.

Valverde, MA; Rojas, P; Amigo, J; Cosmelli, D; Orio, P; Bahamonde, MI; Mann, GE; Vergara, C; Latorre, R. Acute activation of Maxi-K channels (hSlo) by estradiol binding to the beta subunit. *Science.* (1999). *285*(5435):1929-1931.

VanderHorst, VG; Gustafsson, JA; Ulfhake, B. Estrogen receptor-alpha and -beta immunoreactive neurons in the brainstem and spinal cord of male and female mice: Relationships to monoaminergic, cholinergic, and spinal projection systems. *J. Comp. Neurol.* (2005). *488*(2):152-179.

VanderHorst, VG; Terasawa, E; Ralston, HJ 3rd. Estrogen receptor-alpha immunoreactive neurons in the brainstem and spinal cord of the female rhesus monkey: Species-specific characteristics. *Neuroscience.* (2009). *158*(2):798-810.

Vergne, S; Titier, K; Bernard, V; Asselineau, J; Durand, M; Lamothe, V; Potier, M; Perez, P; Demotes-Mainard, J; Chantre, P; Moore, N; Bennetau-Pelissero, C; Sauvant, P. Bioavailability and urinary excretion of isoflavones in humans: Effects of soy-based supplements formulation and equol production. *J. Pharm. Biomed. Anal.* (2007).*43*(4):1488-1494.

Vergne, S; Bennetau-Pelissero, C; Lamothe, V; Chantre, P; Potier, M; Asselineau, J; Perez, P; Durand, M; Moore, N; Sauvant, P. Higher bioavailability of isoflavones after a single ingestion of a soya-based supplement than a soya-based food in young healthy males. *Br. J. Nutr.* (2008). *99*(2):333-344.

Vivacqua, A; Bonofiglio, D; Albanito, L; Madeo, A; Rago, V; Carpino, A; Musti, AM; Picard, D; Andò, S; Maggiolini, M. 17beta-estradiol, genistein, and 4-hydroxytamoxifen induce the proliferation of thyroid cancer cells through the g protein-coupled receptor GPR30. *Mol. Pharmacol.* 2006; 70(4):1414-23.

VonSchassen, C; Fester, L; Prange-Kiel, J; Lohse, C; Huber, C; Böttner, M; Rune, GM. Oestrogen synthesis in the hippocampus: Role in axon outgrowth. *J. Neuroendocrinol.* (2006). *18*(11):847-856.

Wan, Q; Man, HY; Braunton, J; Wang, W; Salter, MW; Becker, L; Wang, YT. Modulation of GABAA receptor function by tyrosine phosphorylation of beta subunits. *J. Neurosci.* (1997). *17*(13):5062-5069.

Wang, CN; Chi, CW; Lin, YL; Chen, CF; Shiao, YJ. The neuroprotective effects of phytoestrogens on amyloid beta protein-induced toxicity are mediated by abrogating the activation of caspase cascade in rat cortical neurons. *J. Biol. Chem.* (2001). *276*(7):5287-5295

Wang, P; Jeng, CJ; Chien, CL; Wang, SM. Signaling mechanisms of daidzein-induced axonal outgrowth in hippocampal neurons. *Biochem. Biophys. Res. Commun.* (2008a). *366*(2):393-400.

Wang, Y; Man Gho, W; Chan, FL; Chen, S; Leung, LK. The red clover (Trifolium pratense) isoflavone biochanin A inhibits aromatase activity and expression. *Br. J. Nutr.* (2008b). *99*(2):303-310.

Waters, EM; Mitterling, K; Spencer, JL; Mazid, S; McEwen, BS; Milner, TA. Estrogen receptor alpha and beta specific agonists regulate expression of synaptic proteins in rat hippocampus. *Brain Res.* (2009). *1290*:1-11.

Watson, CS; Alyea, RA; Jeng, YJ; Kochukov, MY. Nongenomic actions of low concentration estrogens and xenoestrogens on multiple tissues. *Mol. Cell Endocrinol.* (2007). *274*(1-2):1-7.

Wechsler, D. *Wechsler Adult Intelligence Scale-Revised.* Chicago, Illinois: Psychological Corporation, Harcourt Brace Jovanovich, (1981).

Wechsler, D. Wechsler Memory Scale. A standarized memory scale for clinical use. *Psychology*. (1945). *19*:87-95.

Wechsler, D. *Wechsler Memory Scale-Revised*. New York, NY: Psychological Corporation, (1987).

Weiland, NG; Orikasa, C; Hayashi, S; McEwen, BS. Distribution and hormone regulation of estrogen receptor immunoreactive cells in the hippocampus of male and female rats. *J. Comp. Neurol.* (1997). *388*(4):603-612.

White, LR; Petrovitch, H; Ross, GW; Masaki, K; Hardman, J; Nelson, J; Davis, D; Markesbery, W. Brain aging and midlife tofu consumption. *J. Am. Coll Nutr*. (2000). *19*(2):242-255.

Whitechurch, RA; Ng, KT; Sedman, GL. Tyrosine kinase inhibitors impair long-term memory formation in day-old chicks. *Brain Res. Cogn. Brain Res.* (1997). *6*(2):115-120.

Williams, BJ; Bimonte-Nelson, HA; Granholm-Bentley, AC. ERK-mediated NGF signaling in the rat septo-hippocampal pathway diminishes with age. *Psychopharmacology*. (2006). *188*(4):605-618.

Wilson, J; DeFries, J; McLearn, G; Vandenberg, S; Johson, R; Rashad, M. Cognitive abilities: Use of family data as a control to assess sex and age differences in two ethnic groups. *Int. J. Aging Hum. Dev.* (1975). *6*:261-276.

Wise, PM;, Dubal, DB; Wilson, ME; Rau, SW; Böttner, M; Rosewell, KL. Estradiol is a protective factor in the adult and aging brain: Understanding of mechanisms derived from *in vivo* and *in vitro* studies. *Brain Res. Brain Res. Rev.* (2001). *37*(1-3):313-319.

Woclawek-Potocka, I; Piskula, MK; Bah, M; Siemieniuch, MJ; Korzekwa, A; Brzezicka, E; Skarzynski, DJ. Concentrations of isoflavones and their metabolites in the blood of pregnant and non-pregnant heifers fed soy bean. *J. Reprod. Dev.* (2008). *54*(5):358-363.

Wong ,M; Moss, RL. Long-term and short-term electrophysiological effects of estrogen on the synaptic properties of hippocampal CA1 neurons. *J. Neurosci.* (1992). *12*(8):3217-3225.

Wuttke, W; Jarry, H; Becker, T; Schultens, A; Christoffel, V; Gorkow, C; Seidlová-Wuttke, D. Phytoestrogens: Endocrine disrupters or replacement for hormone replacement therapy? *Maturitas*. (2003). *44*(Suppl. 1):S9-S20.

Xiao, CW. Health effects of soy protein and isoflavones in humans. *J. Nutr.* (2008). *138*(6):1244S-1249S.

Xu, X; Wang, HJ; Murphy, PA; Cook, L; Hendrich, S. Daidzein is a more bioavailable soymilk isoflavone than is genistein in adult women. *J. Nutr.* (1994). *124*(6):825-832.

Xu, XW; Shi, C; He, ZQ; Ma, CM; Chen, WH; Shen, YP; Guo, Q; Shen, CJ; Xu, J. Effects of phytoestrogen on mitochondrial structure and function of hippocampal CA1 region of ovariectomized rats. *Cell Mol. Neurobiol.* (2008). *28*(6):875-886.

Yague, JG; Wang, AC; Janssen, WG; Hof, PR; Garcia-Segura, LM; Azcoitia, I; Morrison, JH. Aromatase distribution in the monkey temporal neocortex and hippocampus. *Brain Res.* (2008). *1209*:115-127.

Yanagihara, N; Toyohira, Y; Shinohara, Y. Insights into the pharmacological potential of estrogens and phytoestrogens on catecholamine signaling. *Ann. N. Y. Acad. Sci.* (2008). *1129*:96-104.

Ye, L; Chan, MY; Leung, LK. The soy isoflavone genistein induces estrogen synthesis in an extragonadal pathway. *Mol. Cell Endocrinol.* (2009). *302*(1):73-80.

Zeng, H; Chen, Q; Zhao, B. Genistein ameliorates beta-amyloid peptide (25-35)-induced hippocampal neuronal apoptosis. *Free Radic. Biol. Med.* (2004). *36*(2):180-188.

Zhao, L; Chen, Q; Diaz-Brinton, R. Neuroprotective and neurotrophic efficacy of phytoestrogens in cultured hippocampal neurons. *Exp. Biol. Med.* (Maywood). (2002). *227*(7):509-519.

Zhao, L; O'Neill, K; Diaz-Brinton, R. Selective estrogen receptor modulators (SERMs) for the brain: Current status and remaining challenges for developing NeuroSERMs. *Brain Res. Brain Res. Rev.* (2005). *49*(3):472-493.

Zhao, L; Brinton, RD. Estrogen receptor alpha and beta differentially regulate intracellular Ca(2+) dynamics leading to ERK phosphorylation and estrogen neuroprotection in hippocampal neurons. *Brain Res.* (2007a). *1172*:48-59.

Zhao, L; Brinton, RD. WHI and WHIMS follow-up and human studies of soy isoflavones on cognition. *Expert Rev. Neurother.* (2007b). *7*(11):1549-1564.

Index

A

abolition, 31, 37
absorption, 61
acetylcholine, ix, xi, 49, 68, 73, 74, 77, 83, 89
acid, x, 7, 15, 35, 41, 68, 71
acquisition phase, 60
active treatment, 54, 55
activity level, 28
adaptations, 17, 21, 25, 28
adenosine, ix, 8
adhesion, 19, 71
adipose, 1
adipose tissue, 1
ADP, ix, 8
adrenal gland, 1
adrenal glands, 1
adulthood, 82
aging process, 23, 24
agonist, 9, 20, 71
aldosterone, 7, 85
alters, 88
amplitude, 34, 35, 38, 80
amygdala, 10, 11, 48, 67
amyloid beta, 91
androgen, 7, 26, 48, 81
antisense, ix, 44
anxiety, 69
apoptosis, 43, 48, 81
apoptotic mechanisms, 34
architecture, 12
aryl hydrocarbon receptor, 79
Asia, 49
aspartate, xi
assessment, 51, 55, 57, 73, 81
ATP, ix, 8, 30, 35, 42, 66
avoidance, 28, 48, 76, 77
axon terminals, 11
axons, 79

B

basal forebrain, 18, 27, 49, 70
beneficial effect, 49, 59, 60, 69, 70
bioavailability, 49, 83, 90
biological activity, 79
biological responses, 69
biosynthesis, 71
blood flow, 59
blood-brain barrier, 4
body weight, ix, 35, 47
bone, 3, 80, 83, 86
brain functioning, 54, 58, 61, 62
brain functions, vii, 9
brain structure, 18, 49, 68, 81
brainstem, 67, 90
breast cancer, 43, 44, 82
breeding, 72
buttons, 19

C

calcium, ix, xi, 26, 81, 86
candidates, 67
carcinoma, 42
cardiovascular system, 4
cell culture, 4, 14, 19, 29, 44, 79
cell death, 41
cell line, 43, 88
cell lines, 43, 88
cell membranes, 8, 14
cell surface, 33
central nervous system, vii, 1, 4, 13, 43, 70, 73
cerebellum, 48
cerebral cortex, vii, 10, 48
Chinese women, 57
cholesterol, 1
circulation, 1, 25
class, 1, 5
CNS, 87
cognition, 12, 56, 60, 70, 77, 82, 83, 88, 93
cognitive abilities, 16, 26, 54, 57, 61, 67
cognitive development, 53
cognitive function, 51, 56, 58, 60, 61, 67, 75, 78, 80, 81
cognitive impairment, 87
cognitive performance, 55, 57, 76, 78
cognitive process, 60, 61
cognitive processing, 61
cognitive tasks, 69
cognitive testing, 29
collateral, 31
color, iv
common sense, 69
complexity, 87
composition, 15, 29
compounds, vii, 29, 67
conditioning, 48
conductance, 15, 41, 79
conductor, 9
configuration, 34
configurations, 15, 31, 66, 68
consensus, 8
consolidation, 20, 75

consumption, 37, 49, 55, 69, 75, 91
control group, 59
cooking, 74
copyright, iv
correlation, 54, 62
cortex, 10, 11, 12, 48
cortical neurons, 42, 91
cortisol, 7, 85
critical period, 75
CSF, x, 37, 54
cues, 26
culture, 4, 34, 36, 37, 42
cycles, 62
cyclooxygenase, 48
cytochrome, 18
cytoplasm, 10

D

damages, iv, 33
data availability, 30
deaths, 24
defects, 3, 23, 26, 28, 70
deficiency, 24
dementia, 55
dendrites, 11
dendritic spines, 88
deoxyribonucleic acid, x, 7
dephosphorylation, 33, 74
depolarization, 32, 33, 38
depression, 3, 26
deprivation, 43, 75
detection, 9
developing countries, 88
diet, xi, 37, 45, 47, 52, 55, 56, 58, 59, 61, 62, 69, 84
dietary intake, 54, 58
disability, 24
discrimination, 28, 54
discrimination learning, 54
discrimination tasks, 28
DNA, x, 7, 8, 34, 42
dopamine, 16, 54, 68
dopaminergic, 70
dorsal raphe nuclei, 10, 12

dosing, 36
down-regulation, 19, 25, 48, 69
drugs, 67

E

ecosystem, 23
electrical properties, 31
elongation, 36
emotional state, 69
encoding, 27, 77
endocrine, 1, 4, 25, 68, 72, 81, 85, 89, 92
endocrine glands, 4
endocrine system, 25
entorhinal cortex, 11
environmental change, 23
enzymatic activity, 4, 44, 69
enzymes, 45, 48, 69, 71, 81
episodic memory, 56
ester, 32
estrogen, vii, ix, x, 4, 7, 8, 12, 13, 14, 19,
    20, 26, 31, 44, 49, 59, 60, 65, 67, 69, 71,
    72, 73, 74, 75, 76, 77, 79, 80, 81, 82, 83,
    84, 86, 87, 88, 89, 91, 92
estrogen receptor modulator, 92
ethnic groups, 91
excitability, 27, 28, 37, 38, 39, 67, 69
excitatory synapses, 86
excretion, 2, 61, 90
executive function, 54, 57
executive functions, 54
exons, 7
experimental condition, 33, 39
exploration, viii, 59
exposure, 37, 39, 42, 47, 52, 54, 59, 60, 61,
    62, 69, 87
extinction, 24

F

feedback, 83
fertility, 24
fiber, 18
fibers, 15, 16, 17

flavonoids, 1, 86
flexibility, 54, 56, 61
flora, 2
fluctuations, 25
fluid, x, 37, 54
follicle, 25
forebrain, 49, 70, 74, 84
formula, 53, 61, 77
frontal cortex, 48, 85, 89
frontal lobe, 56, 72, 85
fusion, 26

G

ganglion, x, 14
gene expression, 8, 12, 13, 19, 34, 43, 44
genes, 7, 8, 13, 20, 59
glia, 83
glial cells, 1, 18, 29
glucose, 43
glucoside, 2
glutamate, xi, 17, 32, 34, 43, 68, 71, 73, 79
glutamic acid, x, xi, 5, 13, 15
glycogen, x, 35, 79
gonads, 1
grazing, 3
growth factor, x, 20, 32, 87, 88

H

health effects, vii
hepatocytes, 2
homeostasis, 5
human brain, 49, 78, 79, 87
human cognition, 65
human estrogen receptor, x
human exposure, 47, 49
human subjects, 4, 29, 33, 49, 51, 61, 69
hybridization, 9
hydrolysis, 16, 75
hydroxyl, ix, 1, 32
hydroxyl groups, 1
hygiene, 24
hypothalamus, 10, 11, 12, 45, 47, 86, 87

hypothesis, 75
hypoxia, 43

I

ILAR, 81
immunohistochemistry, 9, 10, 15
immunoreactivity, x, 10, 11, 28, 84
impacts, 18, 30
impairments, 35, 70
in situ hybridization, 9, 10, 11, 15
in utero, 58
in vivo, 3, 91
indirect effect, 86
individualization, 8
induction, 15, 26, 31, 32, 36, 37, 38
industrialized countries, 24
infants, 53, 77
ingestion, 4, 90
inhibition, 32, 37, 38, 42, 43, 44, 73, 84
inhibitor, 16, 19, 30, 31, 32, 35, 43, 73, 85
injections, 35
inositol, xi, 16, 75
insulin, x, 20, 31, 32, 39, 88
integration, 25
intelligence, 55
intelligence quotient, 55
internalization, 32, 39
interneurons, 32, 89
intervention, 37, 54, 60
introns, 7
ion channels, 32
ischemia, 42
isoflavone, 2, 4, 24, 29, 30, 31, 32, 34, 35,
    36, 37, 39, 41, 44, 47, 48, 51, 52, 53, 54,
    55, 58, 59, 61, 65, 66, 67, 69, 76, 77, 78,
    81, 84, 89, 90, 92
isoflavonoids, 1
isolation, 9

K

kidney, 2
kinase activity, 42

L

lactate dehydrogenase, x, 34
latency, 34, 35, 53, 59
learning, 12, 26, 39, 48, 52, 54, 59, 68, 72,
    76, 85
lesions, 85
leucine, xi, 5
life expectancy, 24
ligand, 8, 80
lignans, 1, 54
lipids, 80
localization, 9, 12, 74, 84, 86
locus, 10, 11, 12
long-term memory, 56, 61, 91
LTD, 26, 38, 72
luteinizing hormone, 25

M

magnesium, 72
magnetic resonance, 28, 86, 87
magnetic resonance imaging, 28, 86, 87
magnetic resonance spectroscopy, 28
majority, 9
mammalian brain, 13, 25
manganese, 54, 77
markers, 25
matrix, 2, 62, 67
maze tasks, 28, 58, 76
media, 65
median, 16, 55
medulla, 41
MEK, 44
membranes, 4, 44
memory, vii, 3, 12, 20, 24, 26, 27, 28, 29,
    37, 39, 52, 54, 56, 59, 61, 67, 68, 70, 71,
    74, 75, 76, 77, 78, 81, 82, 85, 91
memory performance, 21, 28, 55, 59
memory processes, 52, 59
menopause, 61, 66, 72, 77
messenger ribonucleic acid, xi, 9
messenger RNA, 89
meta-analysis, 78

metabolism, 4, 34, 44, 49, 61, 73
metabolites, 2, 54, 92
metabolizing, 4, 45, 48, 69, 81
mice, 19, 21, 28, 49, 72, 74, 76, 82, 90
microscope, 30
microsomes, 72
midbrain, 11, 67, 74, 84
mineralocorticoid, 7
mitochondria, 35
mitogen, xi, 14, 15, 79, 81
modification, 61, 74
molecules, vii, 1, 2, 3, 4, 8, 17, 19, 25, 29,
    31, 32, 33, 36, 39, 44, 51, 59, 65, 66, 67,
    70
mood states, 54, 56
morbidity, 24
morphological abnormalities, 35
morphology, 75
mRNA, xi, 4, 8, 9, 10, 11, 17, 18, 19, 20,
    27, 38, 44, 48, 76, 77, 85, 87, 88
mutation, 33

N

neocortex, 92
neonates, 14, 30
nerve, xi, 27, 42, 72, 79, 85
nerve growth factor, xi, 27, 42, 72, 85
nervous system, 73
neuroblastoma, 42
neurodegeneration, 70
neurodegenerative diseases, 24, 70
neuroendocrine system, 25
neurogenesis, 30
neuroimaging, 82
neuronal apoptosis, 34, 92
neuronal cells, 1, 14, 41
neuronal systems, 15, 18, 25, 30, 39, 41, 47,
    66, 70
neurons, 11, 12, 13, 14, 15, 16, 17, 18, 20,
    25, 26, 27, 28, 29, 30, 31, 32, 33, 34, 35,
    36, 37, 38, 39, 41, 45, 48, 65, 66, 68, 70,
    71, 73, 79, 80, 83, 85, 86, 87, 90, 92
neuroprotection, 34, 70, 92
neuropsychological tests, 54, 89

neurotoxicity, 34, 72
neurotransmission, 15, 27
neurotransmitter, 14, 16, 18, 32, 66, 70, 71
nicotine, 39, 74
NMDA receptors, 80, 82, 88
norepinephrine, 16, 68, 75
normal aging, 27, 31
nuclei, 10, 14, 16, 84, 87
nucleotides, 9
nucleus, ix, 10, 11, 12, 16, 48
nutrition, 24

O

oil, 58
oocyte, 43
organ, 4, 65
organic matter, 23
organism, 1, 23
osteoporosis, 3
ovariectomy, xi, 14, 75, 86
ovaries, 1, 20, 25, 60
oxygen, 43

P

parallel, 31, 55, 65
parvalbumin, 19
pastures, 3, 72
pathways, 4, 8, 14, 30, 36, 44, 66
PCR, xi, 9, 10, 11, 15, 27
performance, 20, 25, 28, 52, 54, 56, 58, 59,
    60, 61, 74, 81, 82, 85
performers, 69
permission, iv
phenol, 82
phenotype, vii, 30
phosphorylation, 14, 20, 27, 33, 36, 38, 42,
    66, 74, 81, 87, 89, 90, 92
physiology, 13, 25, 27, 47, 51, 66, 67, 73
placebo, 54, 55, 62, 73
placenta, 83
plants, vii, 2
plasma levels, 25, 61, 67

plasma membrane, 8, 11, 12, 14, 32, 34, 36, 44, 76, 79
plasticity, vii, 25, 71, 76, 81
platform, 53, 60
polarization, 32
polymerase, xi, 9, 88
polymerase chain reaction, xi, 9, 88
pons, 10, 11
population growth, 82
postmenopausal women, 62, 73, 75, 76, 78, 80, 82
potassium, x, 15, 31, 32
prefrontal cortex, 58, 68, 82, 84, 86
prevention, 84
primate, 29, 53, 58, 72
producers, 4
progesterone, 7, 13, 20, 43, 67
pro-inflammatory, 49
proliferation, 42, 43, 90
promoter, 44
protein kinase C, 36, 82
protein structure, 7
protein synthesis, 8
proteins, vii, 5, 7, 8, 9, 13, 14, 17, 19, 20, 27, 48, 52, 65, 73, 74, 79, 81, 91
public health, 24
pyramidal cells, 80

R

radio, ix, 9, 33
reactive oxygen, 34
recall, 13, 54, 56, 61, 65
recalling, 56
receptors, vii, ix, x, 3, 7, 8, 13, 14, 17, 27, 32, 38, 43, 73, 74, 75, 79, 80, 81, 82, 84, 88
recognition, 20, 52, 54, 60, 85
recognition test, 54
recommendations, iv, 24, 66
relevance, 88
replacement, 14, 20, 28, 59, 66, 67, 69, 75, 76, 89, 92
reproduction, 77
reproductive age, 24

residues, 33
resolution, 60
reticulum, 33, 35
reverse transcriptase, xi
ribonucleic acid, xi
rights, iv
RNA, xi, 19
rodents, 26, 29, 76

S

secretion, 25, 85
selective estrogen receptor modulator, 88
semantic memory, 56
sensing, 25
sensitivity, 15
septum, 10, 49, 80
serotonin, ix, 16, 54, 68
sex, 1, 7, 59, 61, 69, 74, 78, 83, 91
sex hormones, 74, 78
sex steroid, 1, 7
shape, 12
sheep, 3, 72, 73
short-term memory, 56, 61
side effects, viii
signal transduction, 28
signaling pathway, 79
signalling, 1, 4, 8, 12, 19, 30, 31, 39, 44, 65, 66, 68, 73, 79, 91
signals, 25, 88
siRNA, 19
sodium, xi, 41, 85
soy bean, 92
soymilk, 92
spatial information, 37
spatial learning, 71
spatial memory, 20, 52, 57, 58, 59, 61, 82, 84, 89
specialization, 24
species, 2, 4, 13, 23, 29, 34, 52, 84, 90
spinal cord, 90
spine, vii, 16, 19, 38, 80, 82
Sprague-Dawley rats, 31, 59
steroids, 14, 83
storage, vii

stressors, 33, 39
striatum, 11, 67
stroke, 87
stromal cells, 44, 75
structural characteristics, 1
substitution, 55, 56
suppression, 71
suprachiasmatic nucleus, 68
surgical intervention, 60
survival, viii, 23, 30, 42, 48, 67, 72
symptoms, 3
synapse, 80
synaptic plasticity, vii, 17, 19, 76, 78, 82, 84, 87
synaptic transmission, 32
synthesis, 1, 16, 18, 19, 26, 78, 80, 86, 87, 90, 92

T

temporal lobe, 85, 86
temporal lobe epilepsy, 85
terminals, 15, 16, 79
testing, 62
testosterone, 1, 18
therapy, 28, 49, 59, 66, 67, 82, 92
thyroid, 43, 90
thyroid cancer, 90
tissue, 4, 9, 80
tofu, 55, 62, 67, 78, 91
toxicity, 91
transcription, 8, 9, 12
transcription factors, 8, 12
transcripts, 10, 11, 27
transduction, 8, 12, 43, 66
transformations, 2, 30
translation, 20

translocation, 14, 32, 36, 37
transmission, 88
transport, 2
trial, 61, 78, 80
trypsin, 67, 71
tryptophan, 13
tyrosine, xi, 30, 31, 32, 34, 38, 42, 43, 48, 71, 72, 73, 74, 80, 81, 84, 85, 86, 87, 90, 91

U

underlying mechanisms, 85
urine, 4, 82, 83, 89

V

validation, 75
vasomotor, 3
vasopressin, ix, 48, 87, 89
vector, 44
vegetables, 2
velocity, 83
ventricle, 36, 37
verbal fluency, 56, 61, 86, 87
victims, 70

W

waste, 37
welfare, 3
western blot, 9, 15, 27
withdrawal, 20, 60, 69
working memory, 18, 20, 54, 57, 60, 61, 75, 85, 89